Distillation Symposium 2022

Topical Conference at the 2022 AIChE Spring Meeting and
18th Global Congress on Process Safety

San Antonio, Texas, USA
10 – 14 April 2022

ISBN: 978-1-7138-6820-0

Printed from e-media with permission by:

Curran Associates, Inc.
57 Morehouse Lane
Red Hook, NY 12571

Some format issues inherent in the e-media version may also appear in this print version.

Printed with permission by Curran Associates, Inc. (2023)

For permission requests, please contact AIChE
at the address below.

AIChE
120 Wall Street, FL 23
New York, NY 10005-4020

Phone: (800) 242-4363
Fax: (203) 775-5177

www.aiche.org

Additional copies of this publication are available from:

Curran Associates, Inc.
57 Morehouse Lane
Red Hook, NY 12571 USA
Phone: 845-758-0400
Fax: 845-758-2633
Email: curran@proceedings.com
Web: www.proceedings.com

TABLE OF CONTENTS

639521 Identifying Energy-Efficient and Heat-Integrated Configurations for Multicomponent Distillation

Abstract:

Multi-component distillation is one of the ubiquitous separation processes found in chemical and petroleum industries. Accounting for about 2.5% of the total energy use in the US [1], distillation is responsible for a significant part of the operating costs and greenhouse gas emissions in a chemical industry. Hence, it is an absolute necessity to explore possibilities by which the energy demand of the distillation configurations can be reduced.

One of the methods by which a reduction in energy consumption can be observed is through recycle of heat between condensers and reboilers. This mode of indirect transmission of heat (without mass transfer) is called as TCH (Thermal coupling via heat transfer) [2]. Since the feasibility and magnitude of the such heat integration links is strongly dependent on operating conditions, it is important that we consider them during the optimization of the configurations. The identification of such 'beneficial' and 'realizable' heat integration links requires a temperature predictor model. One major challenge is that detailed thermodynamic models for temperature prediction create convergence issues in the global optimization framework. In this research we use a simple to use equation, constructed on a theoretical basis, to predict temperatures of non-isothermal phase changing mixtures. We test our model on a few industrially relevant mixtures and validate the same.

Further, we incorporate our model into the global optimization framework to identify configurations with lowest vapor duty (a proxy for energy consumption of the configuration). To ensure finite heat transfer characteristics, we use the concept of MAT (Minimum Approach Temperature) in a heat exchanger. After adding the temperature predictor model and the MAT constraints, we demonstrate the power of our NLP (Non-Linear Programming Framework) by identifying energy-efficient and heat-integrated configurations for four and five component mixtures. These highly energy efficient configurations are simple to implement.

References:

1. S. Environmental Investigation Agency. Monthly Energy Review March 2021.
2. Tony Joseph Mathew, Radhakrishna Tumbalam Gooty, Mohit Tawarmalani, and Rakesh Agrawal. A Simple Criterion for Feasibility of Heat Integration between Distillation Streams Based on Relative Volatilities. Industrial & Engineering Chemistry Research, 60(28): 10286–10302, 2021.

638983 A New Technique to Prevent Minor Component Trapping in Distillation

Abstract:

Minor component "trapping" can result in operational difficulties for the distillation units that are subject to this phenomenon. Difficulties can include products going out of purity specification; surging; and cycling of temperature and composition profiles.

Component trapping can occur in aqueous systems when an impurity that would be normally driven towards the end of a column along with water, forms an azeotrope, which instead drives the impurity towards the other end of the column. In the region of the column where water is absent, the impurity is driven in the normal direction, towards the water-rich section. As a result, the component accumulates in the middle of the column. In continuous units, ppm-level impurities can have mid-column concentration bulges up to tens of percents. When hetero-azeotropes form, a second liquid phase may appear in for several stages in the middle of the column.

A new approach is presented to prevent trapping, for systems where 2 liquid phases appear in a section of a distillation column, e.g. purification of wet polar organics. Partitioning of the components between phases can be exploited to continually remove the trapping-prone minor components while limiting the accumulation and corresponding concentration bulges in the distillation column. By decanting a slipstream from the 2-phase zone, and recycle to single-phase sections, the minor components are continuously removed with the primary distillation products. The results of simulations comparing the concentration profiles / bulges with and without a decanting system are presented.

639508 Accelerating Start-up Using Digital Twin Technology for Combined Virtual Commissioning and Operator Training

Abstract:

Commissioning of process plants is a time-consuming and potentially unpredictable stage on the critical path towards a fully-operating plant. Key technologies for assuring start-up are virtual commissioning and operator training, but these are often disjoint activities, leading to unnecessary expense and investment in model development. This paper presents an approach that streamlines the commissioning and start-up phase while minimizing the total cost of ownership and also results in a digital twin that can be used for multiple further activities over the plant's lifetime.

During the virtual commissioning phase, the automation system design is tested against a relatively simple virtual representation of the process and equipment, which simulates the responses of all devices that communicate with the automation system. Virtual commissioning aims to debug the logic and programming of the control system and Human Machine Interface (HMI) in a virtual environment before commissioning any physical equipment. Correcting even simple bugs can cost orders of magnitude more and add significant delays to start-up.

Similarly, the training of plant operators on the HMI is vital for them to carry out their tasks effectively and safely, particularly on a new process, and to avoid unnecessary shutdowns in the crucial first few months of operation. This training is conducted using an Operator Training Simulator (OTS), which is usually created from scratch as part of the automation system contract.

The process start-up can be significantly streamlined if the virtual commissioning model can be integrated with a full physics-based model of the process. Such an integrated system presents multiple benefits. A consistent look-and-feel of the automation system and OTS means that operators can be trained before the plant is fully commissioned. The use of high-fidelity process models results in an OTS that more accurately represents the dynamic behavior of the plant. Shorter construction time with reduced overall cost can be achieved. Finally, the resulting fully-fledged digital twin can be used for multiple other purposes, such as designing and verifying new control strategies and testing new recipes or revised operating policies.

This work shows how the simulation models developed for testing the automation system can be interfaced with dynamic process models developed and used during process design to provide a high-fidelity OTS. The approach leverages investment in both areas to minimize the incremental cost of creating an OTS while providing a fully-operational digital twin of the process that can create value in other areas.

We describe the workflows of creating such a high-fidelity digital twin, including the challenges and how to address them, using an integrated reaction-and-distillation process as an example. We also describe how the digital twin is a crucial element of a process lifecycle approach that [1] leverages existing investment in R&D and digital design models to reduce start-up time and cost and [2] goes on to utilize such models during operation, in conjunction with real-time plant data, as part of digital operations applications for equipment and process monitoring, soft-sensing, real-time optimization and what-if analysis based on current plant state.

640618 Dual Dividing Wall Columns in Operation

Abstract:

Dual Dividing wall columns (DDWC) is an extension of dividing wall technology with multiple wall arrangements in a single shell. These DDWC combines multiple conventional columns i.e., three or more columns into a single shell and produce four or more high-purity products.

This article shares the operating and design experience of two Dual dividing wall columns operating in Asian refineries.

Light Naphtha Isomerization Unit: The project involved the revamp of a conventional three product Deisohexanizer column to a four product DDWC column. Besides producing a new hexane product, the revamp increased the Isomerization unit capacity by more than 20%.

Lubricating Oil Base Stock (LOBS) Unit: This complex DDWC column operating under vacuum used structured packing as internals. The grassroots column produced four grades of lubricating oil in a single column.

638589 Dr. Izak Nieuwoudt- Mass Transfer Specialist

Abstract:

Dr. Izak Nieuwoudt is a globally known mass transfer expert. His contributions to the world of the separations are immeasurable. Izak's background and experience include:

- Development and testing of mass transfer equipment
- Development of mass transfer processes
- Development of hydraulic and thermodynamic models
- Process simulation and VLE measurements
- Pilot plant design, construction, operation, and EH&S management
- Troubleshooting

Very few chemical engineers have contributed as much to the field of mass transfer as Izak has in academics and industries. Fewer chemical engineers have contributed as much to the field of mass transfer as Izak has fundamentally and practically. This presentation will highlight some of the many contributions of Dr. Nieuwoudt to separations engineering.

Sensitivity of distillation column design to data against which thermodynamic models are correlated

Cara E. Schwarz*

Department of Process Engineering, Stellenbosch University, Banghoekweg, Stellenbosch, 7600, South Africa

* *cschwarz@sun.ac.za, +27 21 8084485*

1. Introduction

The conventional process for distillation column design entails quantifying the separation based on the equilibrium thermodynamics of the system followed by accounting for the degree to which equilibrium is not achieved and the efficiency of the column internals equipment. The first step of this process is dictated by the phase behaviour of the underlying systems and is the focus of this study.

During the design process of a distillation column a thermodynamic model is selected to describe the phase behaviour of the system. This column relates the physical condition in the column to the co-exiting phases and can also provide information on the thermophysical properties of the system. The model is usually selected depending on the system, and the type and degree of non-idealities present and the model parameters are usually fitted to previously measured phase behaviour data data. Both the phase behaviour data as well as the models have inherent inaccuracies. Usually the majority of the inaccuracies of the models are accounted for through fitting the model to the data and safety factors built into the column design. However, uncertainty in the phase behaviour data used to fit the model parameters perturbates to the column design and can exacerbate the uncertainty therein [1] and therefore requires quantification.

Studies have been conducted to determine the effect of the experimental uncertainty of phase behaviour [1–4] and physical property [5] data on the design of distillation columns. These studies have provided an indication of the safety factors that are required due to the uncertainty in the phase behaviour measurements. Other studies have focussed on the uncertainty in the model parameters and determined the influence of perturbation of the model parameters on the distillation column design [6–9].

To the best of the author's knowledge, to date no study has compared the design of distillation columns based on different sets of experimental phase behaviour data available. Typically, during the design process, engineers select what they believe is a suitable set of phase behaviour data, fit their model parameters to this data and base their design on these fitted model parameters. However, for many systems multiple data sets exists at the same or similar conditions and subtle differences between the data sets will propagate to differences in the distillation column design/separation prediction.

The aim of this study is to investigate how sensitive a distillation column design/separation prediction is to the data set which is used to correlate the thermodynamic model parameters.

To achieve this aim, the ethanol (1) + water (2) system was considered and modelled with the well-known Non-Random-Two-Liquid (NRTL) activity coefficient model (ACM) of Renon and Prausnitz [10], as implemented in Aspen Plus ® V11. Both the fitting of the model parameters as well as the simulation of the distillation column was conducted in Aspen Plus.

The ethanol + water system was selected as the system has been well measured with over 220 systems in the NIST ThermoData Engine (TDE) in Aspen Plus® V11. This highly non-ideal system also has significant industrial application as ethanol obtained through the fermentation of biomass is used as a biofuel. The NRTL ACM was selected as it is known to be a suitable model to represent alcohol + water system. Aspen Plus® was selected as it represents a typical process simulator in which such a design task may be fulfilled.

2. Methodology

Overarching, this study was conducted by collecting and evaluating data, fitting the NRTL parameters to the data sets, evaluating the NRTL parameters and the fits obtained and finally simulation of the separation that can be achieved.

2.1. Collection and Evaluation of Data

Phase equilibria data available for ethanol + water systems were obtained from NIST TDE in Aspen Plus V11.0. Each data set was evaluated to determine its inclusion in the study based in the following requirements:

1. All data points should contain values of pressure, temperature, liquid composition and vapour composition. Where one of more of these parameters are not included the data set is omitted.
2. All data points should be at a pressure of no higher than 5 bar. The NRTL model as used in this work assumes that the vapour phase is ideal and is thus only applicable to low pressures. While the 5 bar limit is arbitrary, it allows for a balance of many data sets and adherence to the assumption of an ideal vapour phase.
3. Only data sets with 5 or more mixture data points are included in the study. Further, at least one data point should be at $x_1 > 0.25$ and one at $x_1 > 0.75$ to ensure sufficient compositional range of the data considered.
4. All data sets should pass the Redlich-Kister area test for thermodynamic consistency, as implemented in Aspen Plus® V11 with a limit of 10%. While this tests is not sufficient to ensure accuracy of the data, it is able to eliminate some inaccurate data.

Once suitable data sets were identified, where possible, comparisons were made at constant temperature/pressure to ensure agreement between the various data sets. Data that are not in agreement with other measured data were omitted.

2.2. Fitting of the NRTL parameters to data sets

The expression for the activity coefficient of component i in a mixture, γ_i, of the NRTL model as used in this study is given in equation [1] and the model parameters are defined in Equation [2] to [6]:

$$\ln \gamma_i = \frac{\sum_j x_j \tau_{ji} G_{ji}}{\sum_k x_k G_{ki}} + \sum_j \left(\frac{x_j G_{ij}}{\sum_k x_k G_{kj}} \left(\tau_{ij} - \frac{\sum_m x_m \tau_{mj} G_{mj}}{\sum_k x_k G_{kj}} \right) \right) \quad [1]$$

Where

$$G_{ij} = \exp(-\alpha_{ij} \tau_{ij}) \quad [2]$$

$$\tau_{ij} = a_{ij} + \frac{b_{ij}}{T} \quad [3]$$

$$\alpha_{ij} = c_{ij} \quad [4]$$

$$c_{ij} = c_{ji} \quad [5]$$

$$a_{ii} = a_{jj} = b_{ii} = b_{jj} = 0 \quad [6]$$

For a binary system 5 parameters are required: a_{12}, a_{21}, b_{12}, b_{21} and $c_{12} = c_{21}$. The default parameters available in Aspen Plus® were used as the initial values for the fittings. In this study only a_{12} and a_{21} were fitted. The default values of b_{12}, b_{21} and c were retained. The temperature dependence of τ_{12} and τ_{21} is captured in the values of b_{ij} and b_{ji}. However, for isothermal data sets, no temperature dependence is available in the data and therefore the fitting of b_{ij} and b_{ji} is not possible. Also, due to the close boiling points of the components of the system, the temperature dependence obtained from isobaric data may be limited. Further, the recommended value $c = 0.3$ for alcohol water systems will be retained. Finally, as the number of data points are limited, only fitting a_{12} and a_{21} assists in ensuring that sufficient data are available to achieve the required fit. The a_{12} and a_{21} were fitted using the default maximum likelihood objective function and the Britt-Luecke algorithm.

2.3. AspenPlus® simulation of the separation

Two separation processes were simulated. The first simulation (Column 1) was setup to that pinching close to the ethanol + water azeotrope and a pure water bottoms did not occur and that differences in the simulations are noticeable while at the same time ensuring the simulation is robust and will converge for all cases. The second simulation (Column 2) was setup to be a closer representation of a real separation process and the overheads product may come close to the azeotropic composition.

The RADFRAC unit operation was used with 5 and 20 stages in the column, respectively, the feed on stage 4 and stage 18, respectively, a total condenser and a kettle reboiler. In both cases the condenser pressure was set at 1.0 bar and for simplicity pressure drop across the stages was not considered. Further, both columns were fed with a 1 $kmol.s^{-1}$ saturated liquid feed at 1.0 bar containing 0.15 mole fraction ethanol, the remainder water. The column was specified with a reflux ratio of 0.7 and a distillate withdrawal rate of 0.15 $kmol.s^{-1}$. The simulation was run for the different sets of model parameters to determine their effect on the simulation outputs.

3. Results and Discussion

3.1. Collection and Evaluation of Data, and Fitting of Model Parameters

A total of 225 data sets are available in NIST TDE in Aspen Plus V11. From these 225 data sets 144 data sets were omitted due to incomplete data, incorrect range or too few data points. A further 4 data sets were omitted as they did not pass the Redlich Kister area test for thermodynamic consistency. Comparison of the data at various pressures/temperatures where possible omitted another 6 data sets. It should be noted that, comparison was not possible at all conditions and thus not all erroneous data may have been eliminated. In total 71 data sets will be used for model fitting.

From the 71 data sets identified to be suitable for modelling, 46 data sets provided regression results without warnings or errors. Therefore, together with the default parameter set, 47 sets of parameters were generated.

3.2. Fitting and Evaluation of the NRTL Model Parameters

Evaluation of the parameters shows that there is a strong correlation between the values of a_{12} and a_{21} (See Figure 1). The data can be correlated with a linear relationship with an R^2 coefficient of 0.875. The average values for a_{12} and a_{21} are -0.8322 and 3.517 with standard deviations of 0.1149 and 0.2250, respectively. These values differ from the default parameter values of -0.8007 and 3.458 yet are of the same magnitude. Due to the strong link between the values of a_{12} and a_{21}, further analysis in this work will consider the dependency of various factors only as a function of a_{21}.

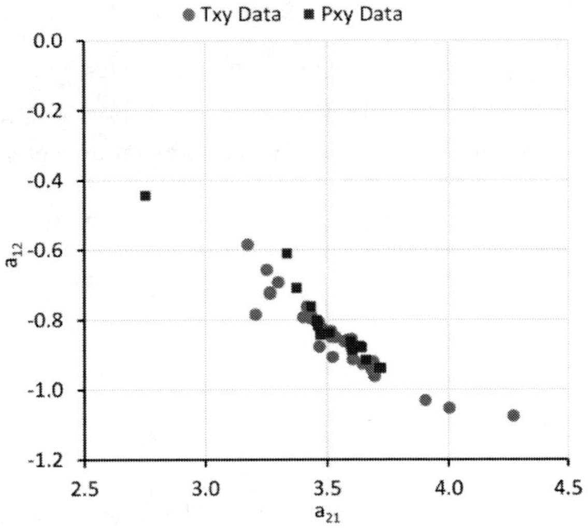

Figure 1: Plot of a_{12} as a function of a_{21} to show intercorrelation of regressed NRTL parameters

Importantly, despite the intercorrelation of a_{12} and a_{21}, Figure 2 shows that the value of a_{21} can not be tied to the system temperature (for isobaric data the saturation temperature of ethanol at the system pressure is considered). Also, as shown in Figure 1 the type of data against which the data are regressed also does not influence the parameters resulting from the correlation. It is therefore postulated that the variation in the parameters may result from subtle differences in the phase equilibrium data. See for example the Txy data at approximately 350 K (101325 Pa) and the variation in the a_{21} parameters associated therewith.

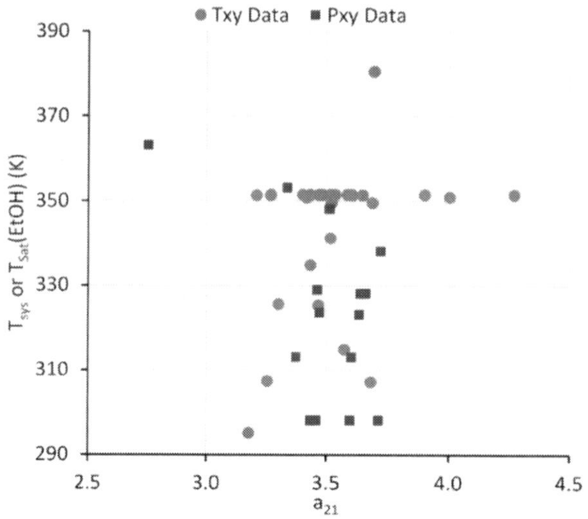

Figure 2: Dependency of regressed a_{ji} parameters as a function of the temperature of data against which they were fitted

The effect of the different model parameters can be seen by considering their effect on the infinite dilution activity coefficient, as shown in Figure 3. As expected, both properties are influenced by the values of a_{21} and therefore it is expected that the separation processes will also be influenced by the model parameters used.

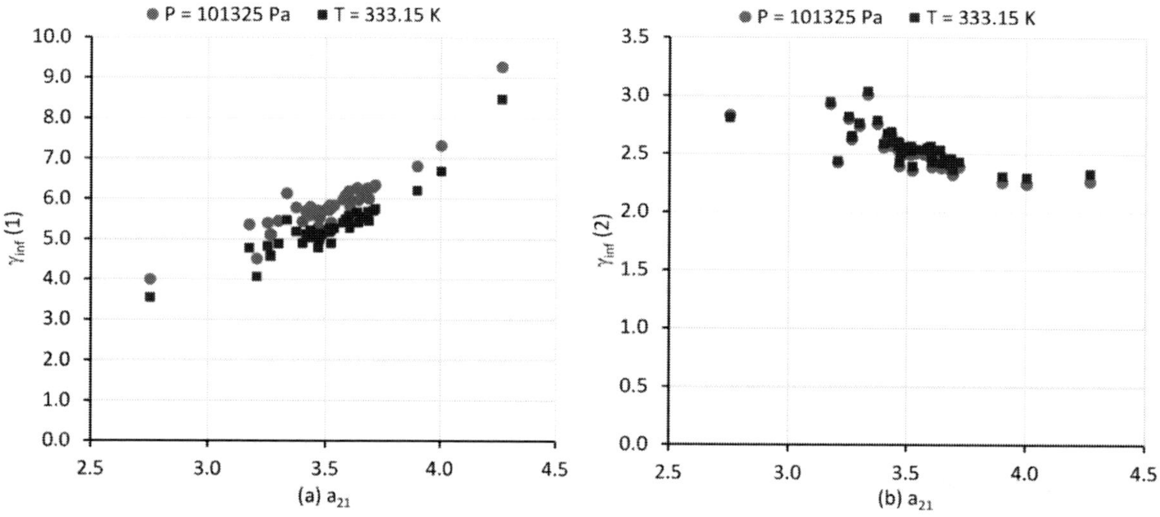

Figure 3: Infinite dilution activity coefficients (a) γ_1^∞ and (b) γ_2^∞ at 333.15 K and 101325 Pa as a function of a_{21}

3.3. Separation Prediction of Various Parameters Sets

The different parameter sets have a noticeable effect on the distillate and the bottoms compositions (See Figure 4). For column 1 there is approximately a 0.04 mole fraction compositional variation in the distillate composition and nearly a 0.01 mole fraction variation in the bottoms composition. Similarly for column 2 there is an 0.06 and 0.01 mole fraction variation in the distillate and bottoms compositions. While a significant amount of the data are clustered around the averages (0.639, 0.0637, 0.807 and 0.0341) the variations may be significant when tight specifications on the composition are required.

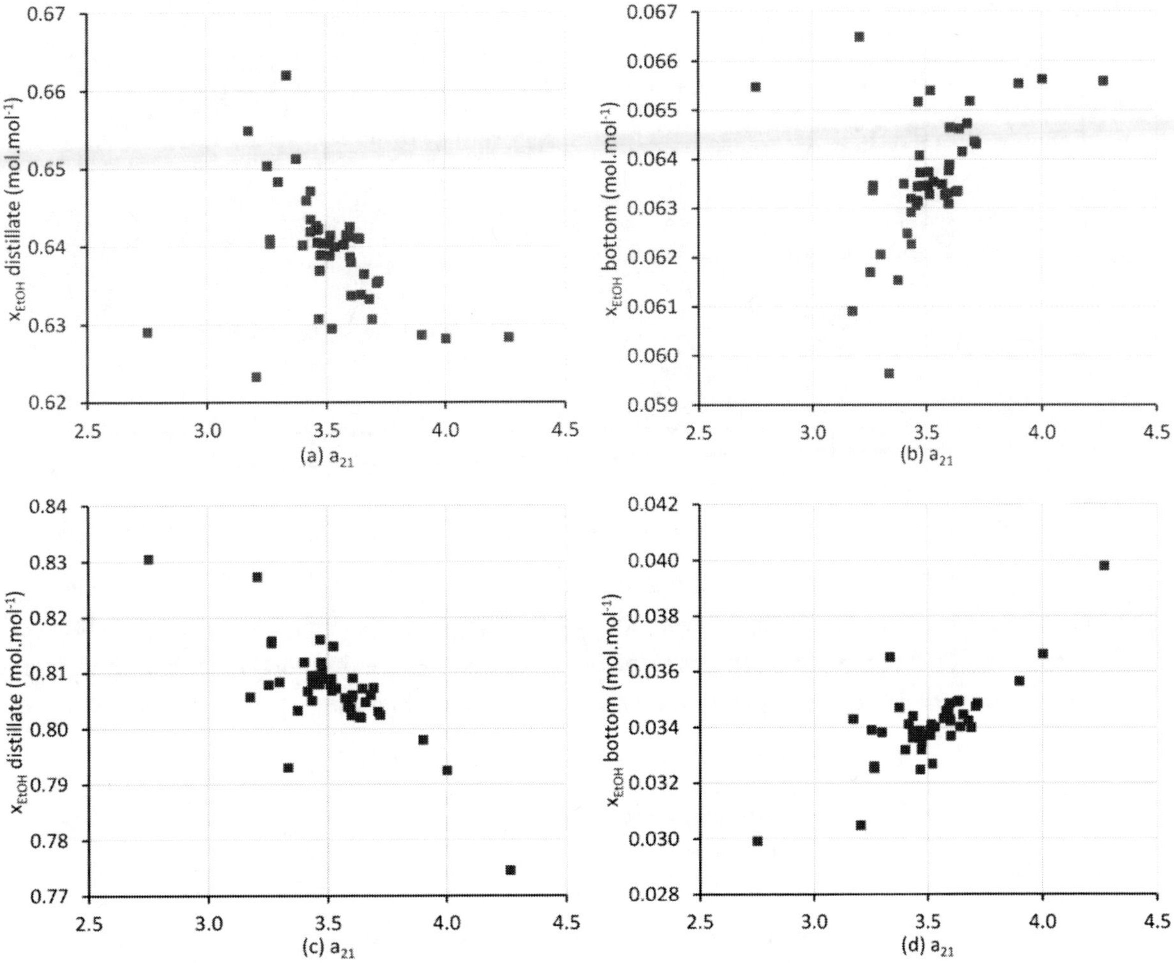

Figure 4: Distillate (a and c) and bottoms (b and d) product composition variation with a_{ij} for column 1 (a and b) and column 2 (c and d)

Similarly, the model parameters strongly influence the condenser (distillate at saturation) and bottoms temperatures (See Figure 5). Therefore, simulations alone can not be used to link temperature and composition in either the distillate or the bottoms of the column.

Further it is noted that for both columns there is a greater variation in the distillate compositions and temperatures than in the bottoms. This could be attributed to the fact that in both cases the feed to the column is closer to the bottom of the column and therefore the greater number of stages above the feed will result in a greater effect of phase behaviour on the separation achieved in the column.

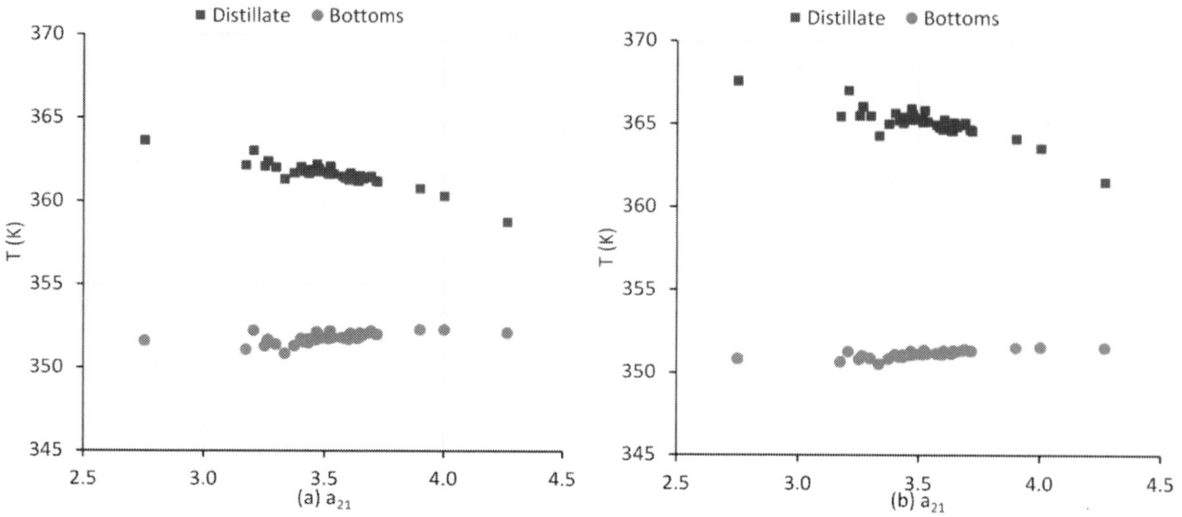

Figure 5: Distillate and bottoms temperature predictions of (a) Column 1 and (b) Column 2 as a function of a_{ji}

Finally, Table 1 shows the variation in the condenser and reboiler duties. It is seen that range of variation is small and therefore the variation in the phase behaviour does not significantly influence the condenser and reboiler duties. The variation noted here is well below 0.05 times of the absolute value and therefore in all likelihood safety factors in the design process will account for these variation.

Table 1: Summary of variation in condenser and reboiler duties

	Column 1 duties (kW)		Column 2 duties (kW)	
	Condenser	Reboiler	Condenser	Reboiler
Minimum	-10164	10702	-14227	15022
Maximum	-10325	10382	-14310	14632
Range	161	321	82	390
Average	-10226	10561	-14262	14861
Standard deviation	23.1	55.5	17.2	65.8

4. Conclusion

This study has shown that the data against which the NRTL model parameters are regressed for the ethanol + water system has a significant influence on the phase behaviour prediction, which in turn propagates to the simulation of a water + ethanol separation column.

Analysis of the model parameters shows that if the temperature dependent of the model are kept at the default values, the a_{12} and a_{21} parameters are strongly intercorrelated with an approximately linear relationship. Further, this correlation propagates to the prediction of the infinite dilution activity coefficients. Unfortunately, no relationship was found between the model parameters obtained and the system conditions. It can thus be postulated that the variation in the model parameters may be ascribed to variation in the experimental data used.

The variation in the model parameters propagates to variation in the distillate and bottoms temperatures and composition with stronger variation being present in the distillate than the bottoms for both temperature and composition. This may be due to the feed position being closer to the bottom of the column. However, the variation on the condenser and reboiler duties are less prominent and will in all likelihood be accounted for safety factors used during design.

In the future the temperature dependence of the model parameters need to be considered as well as others systems with different type of phase behaviour. Further, a comparative study where the effect of experimental uncertainty versus the data set used should be conducted to determine which of these two aspects are more important.

5. References

[1] S. Hajipour, M.A. Satyro, M.W. Foley, , Fluid Phase Equilibria. 364 (2014) 15–30.

[2] L. Burger, C.E. Schwarz, Fluid Phase Equilibria. 458 (2018) 234–242.

[3] W.B. Whiting, J. Chem. Eng. Data. 41 (1996) 935–941.

[4] W.B. Whiting, T.M. Tong, M.E., Ind. Eng. Chem. Res. 32 (1993) 1367–1371.

[5] W.A. Wakeham, G.S. Cholakov, R.P. Stateva, Fluid Phase Equilibria. 185 (2001) 1–12.

[6] P.M. Mathias, H.Z. Kister, J. Chem. Eng. Data. 62 (2017) 2872–2883.

[7] P.M. Mathias, J. Chem. Eng. Data. 61 (2016) 4077–4084.

[8] P.M. Mathias, Fluid Phase Equilibria. 408 (2016) 265–272..

[9] P.M. Mathias, J. Chem. Eng. Data. 59 (2014) 1006–1015..

[10] H. Renon, J.M. Prausnitz, AIChE J. 14 (1968) 135–144.

6. Nomenclature

a_{ij}	Temperature independent contribution to the enthalpic NRTL model parameter
b_{ij}	Temperature dependent $\left(\frac{1}{T}\right)$ contribution to the enthalpic NRTL model parameter
c_{ij}	Entropic NRTL model parameter
G_{ij}	Expression combining the enthalpic and entropic NRTL model parameters
T	Temperature (K)
x_i	Mole fraction of component i in the liquid phase
y_i	Mole fraction of component i in the vapour phase
α_{ij}	Entropic contribution to NRTL model (symmetrical parameters)
γ_i	Activity coefficient of component i in solution
τ_{ij}	Enthalpic contribution to NRTL model (asymmetrical parameter)

639592 Distillation Column Digital Twin with Tower VIEW® Software

Abstract:

TOWER VIEW® software, by Koch-Glitsch and OnPoint, is an innovative digitalization platform, integrated with plant data, and provides real time insights into distillation tower performance. The cloud-based software uses proprietary models from Koch-Glitsch to continuously monitor tower hydraulics and alert the user to impending upsets, flooding, or other hydraulic issues.

This paper shows the implementation of the TOWER VIEW® software digital twin solution for monitoring, optimization, and troubleshooting of distillation towers.

639420 Revamp with Koch-Glitsch PRO® Valve Trays Doubles Run Length on Caustic Absorber

Abstract:

There exist numerous accounts in literature on improvements in run length that can be achieved in distillation or absorption columns in fouling service when replacing sieve trays or movable valve trays with fixed valve trays with large valve openings. In this paper a case study is presented where a revamp with PRO® Valve trays from Koch-Glitsch have been used with good success in severe fouling service.

A newly commissioned olefins plant that catalytically cracked a light Fischer-Tropsch derived olefins stream, was bottlenecked by the run length in the plants' acid gas removal section. An extensive sampling campaign and lab bench scale work identified Aldol condensation polymerisation to be the root cause of the fouling. Extensive research efforts were directed at reducing the reactive components in the feed gas, and optimising the fouling inhibitor dosing chemicals and rates. These resulted in maximum run lengths of between 450 to 600 days on the caustic absorber before it had to be shut down for cleaning. Replacing the original sieve trays in the column with PRO® Valve trays increased run length to more than a 1000 days. This study shows that while it is important to understand the fouling chemistry and the additives that can prevent or reduce the fouling rates, of equal importance is the installation of the correct column internals to minimise the inevitable build-up of fouling material.

638728 Application of Air/Water Derived Structured and Random Packing Mass Transfer Models to Distillation

Abstract:

Application of Air/Water Derived Structured and Random Packing Mass Transfer Models to Distillation

Frank Seibert

Separations Research Program

The University of Texas at Austin

Structured and high performance random packings are often used in distillation, absorption and stripping applications because of their high capacity and mass transfer efficiency along with low pressure drop compared to trayed internals. Packing efficiency is often described and modelled in terms of the volumetric mass transfer coefficient (Ka), the product of the overall mass transfer coefficient (K) and the effective mass transfer area (a). While the packing volumetric mass transfer coefficient is easily measured, the direct measurement of either the mass transfer coefficient or the effective area was a significant challenge. In 2002, the Separations Research Program (SRP) performed their first direct measurement of the effective mass transfer area using a structured packing. These measurements enabled the direct experimental determination of packing gas and liquid film mass transfer measurements. In collaboration with the Rochelle Texas Carbon Management Program with application to carbon dioxide absorption and stripping, the SRP developed an extensive Air/Water related packing database which includes measurements of effective mass transfer area, gas and liquid film mass transfer coefficients. Several UT-Austin Ph.D. students including Tsai (2010), Wang (2015) and Song (2017) developed generalized fundamental models based on the database. A question often asked is whether these Air/Water related models can be used to estimate the packing mass transfer efficiency in hydrocarbon distillation service?

In this presentation honoring Dr. Izak Nieuwoudt, the application of the latest generalized models developed by Song (2017) for structured packing and a new set of models for random packing will be tested with distillation data obtained from an equivalent diameter column. The distillation data are based on the cyclohexane/n-heptane test mixture operated at total reflux and four pressures (2.4, 4.83, 24 and 60 psia). An advantage of the test mixture is the pressure variation allows for changes in the liquid film mass transfer resistance. Both sets of mass transfer models were developed using air/water related test mixtures with small variations in physical properties. The application of the Air/Water derived models will be challenged by the very different and wider range of distillation physical properties. The presentation also provides a methodology for estimating a packing HETP based on the general packing geometry, physical properties and flow data.

Tsai, R., **Mass Transfer Area of Structured Packing**, The University of Texas at Austin, Ph.D. Dissertation 2010.

Wang, C., **Mass Transfer Coefficient and Effective Area of Packing**, The University of Texas at Austin, Ph.D. Dissertation 2015.

Song, D., **Effect of Viscosity on Liquid Film Mass Transfer for Packings**, The University of Texas at Austin, Ph.D. Dissertation 2017.

640209 Adsorption: A New Look at an Old Unit Operation

Abstract:

Adsorption: A New Look at an Old Unit Operation

While benzene is a widely used feedstock in the production of polymers, lubricants, dyes, pharmaceuticals, and pesticides; it is increasingly a cause for concern due to its carcinogenicity. Thus, removing trace amounts of benzene from contaminated chemical inventories becomes an important step in chemical processing. Such contamination may occur in plant storage tanks, transfer lines, and poorly flushed batch and semi-batch process equipment.

Removal of parts-per-million benzene through distillation is costly and often produces product and by-product streams that still contain measurable amounts of benzene. Similarly, liquid extraction could achieve the desired product purity but still involves an expensive distillation step to recover the solvent.

Using an adsorbent to remove benzene from a liquid solution can be both more effective and economical than distillation. However, benzene does not readily adsorb onto activated alumina, and activated carbon will non-selectively adsorb all hydrocarbons. Alternatively, molecular sieves can be used to adsorb benzene by taking advantage of their size exclusion properties.

In this presentation we show thermodynamic data and discuss the implications for the liquid-phase adsorption of benzene from 1-hexene and n-hexane on molecular sieves. Additionally, we discuss future work on the development of a continuous process in the form of a packed bed adsorber as well as modeling and scale-up possibilities.

643662 Evolution of Distillation Equipment Technology and Future Development

Abstract:

Distillation is the most important unit operation in chemical process industry. Both trayed and packed columns are widely used in distillation and absorption applications. Over past 100 years, many new generation distillation trays and packings that exhibit better hydraulic and mass transfer performance had been developed. In this presentation, the evolution of the tray and packing technologies in past 10 years will be reviewed. The latest innovation and development on distillation equipment will be discussed. The areas for future research and development on tray and packing technologies and other column internals will be presented.

639564 "Distillation Sieve Trays without Downcomers: Prediction of Performance Characteristics. Revisited."

Abstract:

In 2002, Garcia and Fair[1] proposed a rational method for the analysis and design of dual-flow tray distillation columns. At 20 years of its publication, there has been only modifications to be able to address specific chemical applications.

Garcia and Fair's method is revisited to present its usefulness to determine operating conditions that can lead to fatigue cracking caused by flow-induced vibrations.

Perhaps the work reported here will help reactivate interest in a device that is efficient, inexpensive and that can operate well under fouling conditions.

1. Antonio Garcia, J. A. & Fair, J. R. "Distillation Sieve Trays without Downcomers: Prediction of Performance Characteristics." *Ind. Eng. Chem. Res.* 2002,41,1632-1640.

639160 Effect of Stripping Factor on Distillation Column Efficiency

Abstract:

Distillation is one of the oldest methods for separating binary and multicomponent liquid mixtures and comprises about 95% of all the separation processes in the chemical and processing industries. Being an energy intensive process, distillation typically contributes up to 50% of capital and operating costs of a plant. Therefore, even small improvements in distillation operation and design would have a significant impact on the costs due to the scale of operation.

The point efficiency plays an important role in estimating tray and column efficiencies during scale up of distillation columns. Most data used to develop efficiency models were obtained from distillation tests under total reflux conditions. However, commercial distillation columns are generally operated in a non-total reflux mode of operation. Therefore, the current efficiency models may not be truly representative of actual efficiencies observed in the field. There is a need to develop reliable efficiency models using data on both total reflux and non-total reflux conditions.

This study aims to fill the gap in the existing literature and provide new experimental data based on non-total reflux mode of operation. This is an extension of our earlier study that collected experimental data based on total reflux mode of operation. As part of this current study, a modified Oldershaw column was designed and installed. The Oldershaw column is capable of operating in both total reflux and non-total reflux modes of operation and provides simultaneous point efficiencies for rectification and stripping sections of the column. Since the stripping factor plays an important role in determining efficiencies in non-total reflux mode, the effect of stripping factor on distillation efficiencies was systematically investigated over several test systems covering a range of physical properties.

639248 Innovations in Pilot Plant Process Distillation Design

Abstract:

MATRIC is a provider of expertise in chemical process development, in process conceptualization, simulation, experimental design and operation. We encounter a wide variety of technologies and problems, depending on our client's needs.

Distillation is a critical element of many chemical processes, and there are numerous problems that require innovative designs. This talk will cover many of the problems we encounter in the course of our work, with the focus being on small scale distillation. Examples of issues that we have encountered are:

- Heat loss in high temperature, small diameter columns
- High vacuum operation
- Compounds with freezing points above ambient
- Accurate measurement of low flow rates
- Cost effective development of model for reactive distillation
- Process sampling
- Development of integrated unit operations
- Product decomposition

We will provide examples of how we tackled these problems. The goal is to provide an approach to how these problems were solved which will have general applicability during the normal work of development.

639262 Reactive Distillation in a Dividing Wall Column

Abstract:

Reactive dividing wall columns (RDWCs) are highly integrated distillation systems that can simultaneously carry out chemical reactions and multicomponent separations within the same vessel. While reactive distillation columns (RDCs) are well established and dividing wall columns are becoming increasingly popular in chemical processes due to their ability to present significant capital and operating costs savings, RDWCs have not been commercially adopted, in part due to the lack of experimental studies and validated models. Therefore, to increase public literature knowledge in this area, a complex aldol condensation reaction was studied at the laboratory scale. The first portion on this work focused on experimentally determining the phase equilibria and kinetics of the test system so that the results could be used to construct a simulation. Next, a simplified RDC was designed, built, and tested with the purpose of using the data to validate phase equilibria and kinetics modeling and inform RDWC design. The RDC experiments and simulation were then used to build, operate, and model a laboratory-scale RDWC. By using the tools of modeling and experimentation to intensify the aldol condensation test system from kinetics and phase equilibria experiments to a RDC then a RDWC, this work contributed a roadmap for evaluating the commercial potential of a RDWC system. Additionally, this work validated the approach of using scaleable Oldershaw laboratory distillation glassware to investigate a RDWC and for model validation.

639590 Composition Proxies in Refinery Distillation

Abstract:

Classical distillation calculations focus on separating specific components. Concentrations of specific chemicals define the product compositions. Refinery distillations have wide boiling ranges and relatively dull separations, considerable amounts molecules with the same boiling point may split between the different products. Due to the number of components, refinery distillation calculations use lumped pseudocomponents defined with vapor pressure and thermodynamic properties defined by standardized methods.

In place of component purities, the refinery uses four major proxies for compositions in fractionation calculations. All of these proxies use a measure of vapor pressure that is not directly a measure of pure component properties. These are:

1. Distillation curves (often called vaporization curves)
2. Reid vapor pressure (RVP)
3. Flash point
4. Noack volatility

Since these proxies are related to vapor pressure, they do have a fundamental link to composition changes. However, they have complex behaviors and use prediction methods that may be based on limited data leading to uncertainties and difficulty in interpretation. This paper examines the inter-relationships between these purity proxies and how calculation methods and correlation accuracy can affect analysis and design of refinery distillation units.

640352 Surface Tension and Distillation Design

Abstract:

Surface tension is one of the key physical properties commonly used in design correlations for mass transfer and capacity of separation equipment, because of its effects on droplet size, film behavior and vapor-liquid disengagement. Specifically, surface tension is used in predicting mass transfer efficiency of trays and packings, entrainment behavior, downcomer performance as well as system limit of distillation columns. Mass transfer and design correlations typically use surface tension measured by observing the behavior of a liquid droplet in air at atmospheric pressure, as a surrogate for the interfacial tension of a liquid in contact with condensable vapor. Correlations used in distillation design and rating have used this approach for decades, simply because surface tension data measured in air at ambient conditions is more readily available. This presentation will discuss the differences between surface tension measured in air versus a condensable vapor and potential ramifications on the physical understanding of distillation processes.

639117 Batch Distillation Robust Design and Operation in the Face of Feed Variability and Thermodynamic Uncertainties Using Stochastic Modeling in Multibatchds™

Abstract:

Separations using batch distillation often encounter the problem of feed variability. One of the significant problems in optimizing or simulating batch distillation operation for a particular product specification in specialty chemical industries is that the number of components whose thermodynamic properties are known is less than the total number of chemicals in the mixture. In pharmaceutical industries, new drugs bring in new components whose thermodynamic data is restricted to the structure of the molecule. In such conditions, stochastic modeling can be used to study the effect of these uncertainties on product specification and find robust design and operating conditions.

The stochastic modeling involves:

1. Specifying the uncertainties in key input parameters in terms of probability distributions;
2. Establishing the correlation structure of any interdependent parameters;
3. Sampling the distributions of the selected parameters iteratively;
4. Propagating the effects of uncertainties through batch distillation simulation model; and
5. Applying statistical techniques to analyze the effects of uncertainties on the product yield and specification.

Multibatch distillation systems (**MultiBatchDS™**) is a software package that captures the essence of these multiple configurations, multiple operating modes, and various fractions of batch distillation by providing the capability to simulate, design, optimize, and control such columns with multiple levels of models. This paper presents two real-world case studies of feed variability and/or thermodynamic uncertainties in batch distillation using the novel stochastic modeling framework in **MultiBatchDS.™**

640164 Comprehensive Study of Axial Dispersion on Several Modern Structured and Random Packings

Abstract:

Axial dispersion is an undesirable phenomenon and it is not the subject of attention in the literature in the last years. Axial dispersion degrades efficiency of the separation equipment through reduction of the mass-transfer driving force, and in the limit, a column could behave like a single equilibrium stage. This work is focused on determination of axial dispersion in the form of Bodenstein number for liquid phase for five types of modern packings in a packed absorption column. Packings used in this work are random packings NeXRing0.7 and NeXRing1.2 and structured packings MellaPak350Y, 452Y and 500Y. The data obtained from dynamic experiments in the form of a response to a step change in tracer concentration, are compared to diffusion-type differential equation, which was solved with boundary conditions for a closed system. By measuring responses at two different packed bed heights and using convolution, the influence of end effects was eliminated.

The lowest axial dispersion was obtained for M452Y, and it is almost twice as lower than for M250Y. Other structured packings have an axial dispersion comparable, especially at low liquid rates. It was also found that the values of the Bodenstein number for M350Y and M452Y are dependent on the gas flow rate which is in contrary to the literature.

Axial dispersion found for random packings is generally higher compared to structured packings. However, compared to other types of random packings (Pall rings 25mm, Intalox saddle 25mm), NXR0.7 and NXR1.2 have an axial dispersion up to twice as lower.

The power function fits the obtained data for M250Y, M350Y and M500Y with an accuracy of ± 20 %. The correlation for random packings fits the data with an accuracy of ± 15 % for NXR0.7 and ± 20 % for NXR1.2.

639259 The Unexpected Weirdness of Rate-Based Simulation of Some High-Pressure Column Operations

Abstract:

Diffusion is the process by which molecules randomly move towards a phase interface. For the most part, molecules move from a region of high concentration to one with low concentration. It has long been known that diffusion in multicomponent systems is complicated and, in fact, molecules can diffuse from low concentration to high.

One way to model multicomponent diffusion is with a version of Fick's law: (J) = - [D] d(c)/dz where [D] is a matrix of multicomponent diffusion coefficients. [D] is the product of two other matrices, one of which involves partial derivatives of the fugacity or activity coefficients.

Thermodynamics requires that that the matrix [D] be positive definite and it is usually the case that all of the elements on the main diagonal of [D] have a positive value (this is not a requirement of positive definiteness). Ten years ago, Wambui-Mutoru and Firoozabadi showed that CO_2-rich and N_2-rich hydrocarbon systems could have main diagonal elements of [D] that are negative.

Using rate-based column simulations, we have found that these negative main diffusion coefficients readily occur in some high-pressure distillation operations. Our purpose in this paper is to investigate the thermodynamics of these processes in more detail, to identify the circumstances in which negative main diffusion coefficients can occur and to explore the consequences for the design and operation of columns in which this phenomenon is predicted to occur.

639509 Using Thermography for Visualisation of Liquid Surface Activity in Mass-Transfer Processes

Abstract:

Certain gas-liquid mass-transfer processes, such as chemisorption, evaporation, and especially binary distillation, are connected with significant interfacial heat transfer. For these processes, thermography offers a unique opportunity to study the local intensity of the mass transfer via noninvasive measurement of the interface temperature. The thermal imaging of the liquid surface allows us to identify, for example, the constructions effective in preventing the evaporation from water reservoirs. However, it should be also able to reveal the geometrical features of the packing, which promote the mass transfer, or distinguish the liquid phase areas active in interface mass transfer from the 'dead' ones. The wet surfaces with limited mass exchange with the surrounding liquid are brought into equilibrium with the vapor, and such surfaces, yet wet, become inactive in the interfacial mass exchange. The local temperature of these inactive liquid spots rises to the temperature of the vapor, which shall be detected by the thermal camera. In other words, the procedure could evolve into the method for a determination of the effective mass-transfer area under distillation conditions, which is so far inaccessible by experimental techniques.

In this contribution, we present the results of the thermographic liquid surface temperature measurement under two simplified process conditions: (i) water evaporation from a still liquid level in the agitated vessel, (ii) evaporation of a binary system with a laminar flow of the liquid and vapor phase and also (iii) first results with distillation of a binary system with liquid flowing down the inclined plane and vapor flowing counter currently above it.

The results of the still level experiments clearly show that the liquid surface thermal images correspond to the temperature of the gas-liquid interface measured indirectly via the water vapor partial pressure in the gas phase. The data measured in a wide range of experimental conditions confirms that the bulk temperature of the gas does not influence the thermal images and the liquid bulk affects thermograms very weakly.

The surface compositions calculated from the thermal images acquired under the evaporation conditions were compared to those measured via confocal Raman spectroscopy and with the theoretical compositions calculated from fundamental transport relations. The results confirm the possibility of thermographic determination of the liquid surface composition under evaporation conditions, and thus the ability of thermography to analyze the local mass transfer activity of the liquid surfaces.

The distillation experiments will be performed in the following arrangement: the liquid and vapor of a binary system are brought into contact in the rectangular distillation space, where the liquid flows down the inclined plate while the vapor flows above the film counter currently. The liquid film is observable through the heated KBr windows (transparent in IR) built into the lid covering the vapor space. The IR camera monitors the interface temperature field.

We aim to put the occurrence of such spots into context with the system properties (surface tension dependence on the mixture composition). The local packing geometry effect is planned to be studied later, when the experimental apparatus is equipped with a real structure packing sheet.

640114 Liquid Cracker Quench Water Tower Choice of Internals and Operation

Abstract:

Quench tower in a liquid based olefins plant refers to both Quench Oil tower and Quénch Water tower. Gas cracker may have only Quench Water towers. Generally, the Quench Oil tower is subjected to heavy fouling tendency depending on the nature of feed used for cracking. However, the Quench Water tower has much lower potential for fouling in liquid cracker. Large number of Quench Water towers in liquid crackers that are in operation worldwide today are equipped with random packing. While generally agreed that random packing is suitable to handle the fouling tendency, another option is available when studying the problem in detail. In this paper, a different design concept is introduced, which uses structured packings to achieve a great success for the Quench Water tower in liquid crackers for more than 10 years in operation. Based on the recent site inspection, the original structured packings are still in good working conditions to achieve its heat transfer requirement and operable without any fluctuation in the pressure drop. No replacement of either the structured packings or the liquid distributors and other column internals were needed to-date. Besides using the right choice of tower internals, operational excellence which includes running and maintaining the tower by the end-user with the highest standards and broad experience in the Olefins plant operation determines the success of the Quench tower.

640273 Distillation Performance of RMP SP 250 High-Capacity Structured Packing at the FRI Test Facility

Abstract:

RVT Process Equipment´s development of a high-capacity structured packing family was launched in 2014 with the RMP SP 250 structured packing. Distillation tests ran at the Separation Research Program (SRP) pilot plant (ID428 mm) with the C6/C7 system at that time showed excellent performance compared to state-of-the-art high-capacity packings with the same surface area. For all tested operating pressures: 0.165, 0.33, 1.65 and 4.14 bara, RMP SP 250 achieved a remarkable reduction in pressure drop accompanied by the consequent increase in capacity while delivering the expected mass transfer efficiency level (HETP) typical for 250 m2/m3 high-capacity structured packings [1].

In the light of these very promising results, the FRI membership chose the RMP SP 250 for the Category 1 packing in 2020 using FRI´s 1.22 m diameter industrial size column. RVT proposed testing the packing with the C6/C7 system and the two related standard pressures of 0.31 and 1.62 bara, as well as the ortho/para-Xylene system operated at 0.1 and 1.0 bara. In order to save unit time loss related to the liquid distributor change required when shifting within both test systems, RVT designed and delivered a multistage trough type distributor capable of working with the required extremely wide reflux loads, ranging from 2.5 m3/m2h to 40 m3/m2h (6.5-100%).

The distillation measurements ran with RMP SP 250 at FRI show very good performance results for both systems and operating pressures. Mass transfer efficiency levels achieved with both C6/C7 and o/p-Xylene system are in good agreement with data obtained with state-of-the-art high-capacity 250 m2/m3 packings at the same facility. On the other side, specific pressure drop proved to be again significantly lower which allowed for significantly higher hydraulic capacities in both systems and operating pressures. These two key performance parameters combined in the form of the pressure drop per theoretical stage show that RVT´s RMP SP 250 packing clearly delivers the best (lowest) values of all 250 m2/m3 high-capacity packings tested at FRI.

The total reflux distillation measurements ran with the RMP SP 250 at FRI´s industrial column confirm the excellent performance results previously obtained at SRP pilot plant as well as already available practical on-site experience in different industrial columns in which the RMP SP 250 has already been installed.

[1] "Distillation performance characteristics of RMP-type standard and high capacity structured packings",
2016 AIChE Spring Meeting, Advances in Distillation Packing Technology, Houston (TX, USA)

639541 Why NOT Structured Packings in High-Pressure Distillation Applications?

Abstract:

In the chemical processing industries (CPI), structured packings are widely used for distillation and absorption columns. However, the HETP 'hump', or deterioration of the mass transfer performance at about 65-90% of flood, made the structured packing not suitable for high-pressure distillation applications. It is very crucial to understand why the structured packing is not trusted in the high-pressure distillation applications. In this paper, the hump phenomenon and observations are described. The previous studies on the HETP hump are reviewed. The possible explanations of the hump mechanisms are discussed. Based on this study, the likely reasons that may cause the HETP hump are proposed, and a guideline about the structured packing application is provided. Future studies on HETP hump are discussed to further under the hump phenomenon.

639212 Performance Comparison of HETP Correlations for High-Capacity Structured Packings

Abstract:

Previous performance studies on HETP correlations have been conducted using structured packings with relatively low nominal packing areas. Further, the previous studies generally did not include high-capacity structured packings with new geometrical features such as MellapakPlus 252Y, 452Y, and 752Y. This is in part due to the lack of publicly available higher capacity structured packing test data. Therefore, many of the existing HETP correlations in the literature are developed using experimental data with structured packings of low nominal areas. In this study, a performance comparison and comprehensive evaluation of several HETP correlations from the literature will be presented. The HETP correlations from the literature selected for these evaluations include Gualito (1997), Billet and Schultes (1999), Delft (2004), and Hanley (2012). The performance of these correlations was evaluated on experimental HETP data across the pre-loading regime. Experimental HETP data collected for this study represented pressures ranging from 0.10 – 1.62 bar. Both aqueous and non-aqueous test systems were included in the overall evaluation. An analysis of the results and observed strengths and weaknesses of existing literature models in predicting HETP for high-capacity structured packings will be presented.

638945 Progress Towards a Comprehensive Model for Mass Transfer and Hydraulics for Structured Packings

Abstract:

Rate-based calculations theoretically offer more rigor and reliability for assessing column performance than the traditional equilibrium-stage approach. Unfortunately the rate-based method is ultimately tied to equipment performance correlations with limited predictive capability. We have undertaken an effort to develop significantly improved correlating expressions for mass transfer in columns containing sheet metal structured packings. To accomplish this, we have collected and simultaneously analyzed HETP data from 443 distillation experiments (Bennett, et al., 1992; Bravo, et al., 1985; Fitz, et al., 1999; Gualito, et al., 1997; and others) and 410 mass transfer area experiments (Tsai, 2010). The correlating expressions for the mass transfer performance for these sheet metal structured packings are shown below:

$Sh_V = 0.333\ Re_V^{1}\ Sc_V^{1}\ [\cos(\theta)/\cos(\pi/4)]^{0.552}$

$Sh_L = 0.1666 Re_L^{0.8}\ Sc_L^{0.758}\ [\cos(\theta)/\cos(\pi/4)]^{0.747}$

$a_m/a_d = 0.8093\ Re_V^{-0.0293}\ We_L^{0.175}\ Fr_L^{-0.1208}$

$Sh_V = (k_y d_e/(c_V D_V))$

$Sh_L = (k_x d_e/(c_L D_L))$

$d_e = 4\varepsilon/a_d$

The table below compares HETP predictions from the correlation above and from other popular mass transfer correlations in the open literature (Bravo, et al., 1985; Hanley and Chen, 2014; Wang, et al., 2016) with some of the experimental data we have collected. We find that the correlation of this work predicts the experimental HETP dataset with an average difference of 9.4% and a maximum difference of 50%; the next best correlation - that of Wang, et al. - predicts these same data with an average difference of 20% and a maximum difference of 172%.

		Pressure	Packing	Fs	HETP	This Work	Hanley & Chen	Wang, et al.	BRF85
Light Key	**Heavy Key**	**Pa**		**$Pa^{1/2}$**	**m**	**m**	**m**	**m**	**m**
P-XYLENE	O-XYLENE	2000.0	Mellapak 250Y	0.8478	0.3615	0.4016	0.4318	0.4002	0.2549
P-XYLENE	O-XYLENE	13000.0	Mellapak	0.2332	0.3493	0.3587	0.3306	0.2851	0.2154

			250Y						
P-XYLENE	O-XYLENE	100000.0	Mellapak 250Y	0.2817	0.3221	0.3195	0.2677	0.2463	0.2303
IC4	NC4	690000.0	Mellapak 250Y	1.1402	0.2868	0.3061	0.2346	0.3007	0.3002
IC4	NC4	1140000.0	Mellapak 250Y	0.7191	0.2778	0.2773	0.2112	0.2525	0.2131
CYCLOC6	NC7	33000.0	Mellapak 250Y	3.1962	0.4111	0.4080	0.4068	0.4906	0.5029
CHLORBZ	ETHBZ	96000.0	Mellapak 250Y	1.5976	0.3548	0.3387	0.2698	0.3305	0.3636
CHLORBZ	ETHBZ	96000.0	Mellapak 350Y	2.0988	0.2631	0.2571	0.2204	0.2335	0.2671
CHLORBZ	ETHBZ	5000.0	Mellapak 500Y	1.0143	0.2195	0.2141	0.2443	0.1694	0.1196
2ME2BUTL	ISOBUTL	20000.0	Mellapak 500Y	1.2422	0.2805	0.2364	0.4578	0.2673	0.1584
2ME2BUTL	ISOBUTL	10000.0	Mellapak 500Y	1.1324	0.2694	0.2502	0.6447	0.3234	0.1488
2ME2BUTL	ISOBUTL	2000.0	Mellapak 500Y	1.6408	0.3365	0.2934	1.6509	0.5715	0.1538
CHLORBZ	ETHBZ	2000.0	Mellapak 500X	1.3943	0.2846	0.2811	0.7395	0.2240	0.1389
CHLORBZ	ETHBZ	10000.0	Mellapak 500X	1.3977	0.2780	0.2630	0.4882	0.1957	0.1578
CHLORBZ	ETHBZ	20000.0	Mellapak 500X	0.6351	0.2670	0.2490	0.4033	0.1654	0.1365

CHLORBZ	ETHBZ	95000.0	Mellapak 500X	1.9088	0.2434	0.2353	0.2651	0.1741	0.1931
2ME2BUTL	ISOBUTL	10000.0	Mellapak 500X	1.7589	0.3269	0.3225	2.2484	0.3967	0.1927
2ME2BUTL	ISOBUTL	20000.0	Mellapak 500X	1.3913	0.3035	0.2999	1.4575	0.3175	0.1865
2ME2BUTL	ISOBUTL	95000.0	Mellapak 500X	1.4372	0.2582	0.2609	0.5831	0.2208	0.1991
CHLORBZ	ETHBZ	10265.8	Mellapak 250Y	1.4602	0.3889	0.3810	0.3434	0.3772	0.3189
CHLORBZ	ETHBZ	10265.8	Mellapak 250Y	2.4386	0.3975	0.3883	0.3448	0.4072	0.3638
CHLORBZ	ETHBZ	10265.8	Mellapak 252Y	3.0246	0.4071	0.3918	0.3454	0.4208	0.3852
P-XYLENE	O-XYLENE	13332.2	Mellapak 252Y	2.9485	0.3553	0.3861	0.3363	0.4136	0.4035
CYCLOC6	NC7	165474.2	Mellapak 252Y	1.4631	0.3387	0.3479	0.3027	0.3624	0.3981
CYCLOC6	NC7	101325.0	Mellapak 752Y	1.7872	0.2271	0.2028	0.2164	0.1769	0.1943
ISOC8	TOLUENE	13332.2	ISP IT	1.4225	0.3105	0.3282	0.3567	0.3517	
ISOC8	TOLUENE	13332.2	ISP 4T	1.3958	0.5555	0.6257	0.5648	0.8315	
ARGON	OXYGEN	206842.7	Flexipac 500Y	0.9095	0.1811	0.1784	0.1586	0.1523	0.1533
P-XYLENE	O-XYLENE	5332.9	Flexipac 1.4X	2.5618	0.3454	0.3787	0.7143	0.3428	0.2802
			Flexipac						

P-XYLENE	O-XYLENE	26664.5	1.4X	2.2568	0.3493	0.3490	0.4606	0.3011	0.3037
P-XYLENE	O-XYLENE	159986.8	Flexipac 1.4X	1.5127	0.3150	0.3068	0.2799	0.2503	0.2824
P-XYLENE	O-XYLENE	439963.8	Flexipac 1.4X	1.4517	0.2731	0.2801	0.2105	0.2294	0.2549
P-XYLENE	O-XYLENE	5332.9	Flexipac 1.6X	1.2931	0.3531	0.4332	0.7964	0.3858	
P-XYLENE	O-XYLENE	26664.5	Flexipac 1.6X	1.6225	0.3835	0.4043	0.5150	0.3551	
P-XYLENE	O-XYLENE	159986.8	Flexipac 1.6X	1.1101	0.3581	0.3559	0.3131	0.2956	
P-XYLENE	O-XYLENE	439963.8	Flexipac 1.6X	1.0491	0.3302	0.3247	0.2354	0.2699	

In addition, we have used this new correlation to model the performance of several carbon capture pilot plant experiments reported in the literature (Gabrielson 2007 ; Notz et al., 2012) . Those results, along with results for the Bravo, Rocha Fair correlation of 1985 (Bravo, et al., 1985), the 2014 correlation of Hanley and Chen, and the 2016 correlation of Wang, et al., are summarized below. Again, the correlation of this work outperforms the other correlations examined.

Gabrielson Data

Run 1

	Experimental	This work	Wang	Hanley/Chen	BRF 85
Gas CO2 conc. Bottom (%vol)	2.62	2.62	2.62	2.62	2.62
Gas CO2 conc. Top (%vol)	1.69	1.645	1.602	1.089	1.278
Liquid CO2 loading top	0.072	0.07201	0.7201	0.07201	0.07201
Liquid CO2 loading bottom	0.178	0.183	0.1878	0.2451	0.224

Run 3

	Experimental	This work	Wang	Hanley/Chen	BRF 85
Gas CO2 conc. Bottom (%vol)	4.17	4.17	4.17	4.17	4.17
Gas CO2 conc. Top (%vol)	2.36	2.173	2.095	1.514	1.64

Liquid CO2 loading top	0.084	0.084	0.084	0.084	0.084
Liquid CO2 loading bottom	0.169	0.1789	0.1825	0.2093	0.2035

Run 6

	Experimental	This work	Wang	Hanley/Chen	BRF 85
Gas CO2 conc. Bottom (%vol)	4.58	4.58	4.58	4.58	4.58
Gas CO2 conc. Top (%vol)	2.9	2.883	2.785	2.142	2.337
Liquid CO2 loading top	0.147	0.147	0.147	0.147	0.147
Liquid CO2 loading bottom	0.226	0.2272	0.2318	0.2611	0.2523

Run 11

	Experimental	This work	Wang	Hanley/Chen	BRF 85
Gas CO2 conc. Bottom (%vol)	10.27	10.27	10.27	10.27	10.27
Gas CO2 conc. Top (%vol)	8.01	8.048	7.797	6.474	7.108
Liquid CO2 loading top	0.284	0.284	0.284	0.284	0.284
Liquid CO2 loading bottom	0.4	0.3945	0.4057	0.4663	0.438

Notz, et al., Experiment 2 Data

	Experimental	BRF85	Hanley/Chen	Wang	This work
GASOUT mflow (kg/hr)	65.6	66.3391	66.3941	66.6300	66.6127
GASOUT mfrac CO2	0.088	0.0750	0.0757	0.0795	0.0816
LEANIN mflow (kg/hr)	200	198.3004	198.3799	197.6807	197.6336
LEANIN molefrac (nCO2/nMEA)	0.308	0.2988	0.3022	0.2926	0.2899
RICHOUT mflow (kg/hr)	207.4	205.8422	205.9266	205.2253	205.1806
RICHOUT molefrac (nCO2/nMEA)	0.464	0.4692	0.4712	0.4553	0.4490
CO2OUT mflow (kg/hr)	6.14	7.0062	6.9543	6.6847	6.5384
CO2OUT mfrac CO2	0.996	0.9945	0.9945	0.9945	0.9945
WATERMU mflow (kg/hr)	1.21	1.4846	1.5416	1.7765	1.7660
RICHOUT HeatX duty (kW)	12.4036	11.9129	11.8712	11.9612	11.9935
LEANIN H2O (w/w)	0.653	0.6519	0.6514	0.6524	0.6526
LEANIN MEA (w/w)	0.284	0.2864	0.2863	0.2871	0.2873
LEANIN CO2 (w/w)	0.063	0.0617	0.0623	0.0605	0.0600

Future work will attempt to incorporate improved column hydraulic and pressure drop predictions to supplement the mass transfer correlation reported in this talk.

REFERENCES

Agrawal R, Woodward DW, Ludwig KA, Bennett DL. Impact of low pressure drop structure packing on air distillation. Paper presented at Distillation and Absorption: IChemE Symposium Series 128. Maastricht, The Netherlands, 1992.

Bravo JL, Rocha JA, Fair JR. Mass transfer in gauze packings. Hydrocarbon Process. 1985;64:91.

Chao Wang, Di Song, Frank A. Seibert, and Gary T. Rochelle. Dimensionless Models for Predicting the Effective Area, Liquid-Film, and Gas-Film Mass-Transfer Coefficients of Packing. Industrial & Engineering Chemistry Research. 2016 55 (18), 5373-5384

Fitz CW, Kunesh JG, Shariat A. Performance of structured packing in a commercial-scale column at pressures of 0.02–27.6 bar. Ind Eng Chem Res. 1999;38:512–518.

Gabrielsen J. CO2 capture from coal fired power plants. PhD Thesis, Technical University of Denmark, 2007.

Gualito JJ, Cerino FJ, Cardenas JC, Rocha JA. Design method for distillation columns filled with metallic, ceramic, or plastic structured packings. Ind Eng Chem Res. 1997;36:1747–1757.

Hanley B, Chen C-C. Letter to the Editor. AIChE J. 2012;58:132–152.

Notz, R., Mangalapally, H.P., Hoch, S., Hasse, H., 2011. Post combustion CO2 capture by reactive absorption: Pilot plant description and results of systematic studies with MEA. International Journal of Greenhouse Gas Control 6 (2012), 84-122.

Tsai, Robert E. Mass Transfer Area of Structured Packing. Ph.D. Dissertation. 2010.

639505 Mass-Transfer Model for Distillation Columns Equipped with Structured Packings

Abstract:

The contemporary mass-transfer models are mostly verified by the *HETP* values obtained with standard distillation systems. However, the analysis of the composition profiles along the distillation column packed bed allows us more detailed analysis of the mass-transfer and hydrodynamic phenomena. It enables us to quantify distribution of the mass-transfer resistances and deviation from the plug flow of the phases.

In this work we present a progress in the development of the advanced mass-transfer model for structured packings, which is based on: a) analysis of the distillation composition profiles in the industrially accepted size of the equipment (providing $k_L a$, $k_G a$ instead of *HETP*); b) usage of the non-ideal hydrodynamic model for the liquid phase (plug flow with the axial dispersion); c) utilization of the data acquired with proven absorption test methods with non-air (k_G) and non-aqueous (k_L) systems and with more than one transferring solute; d) incorporation of the Marangoni effects influence in the model for the interfacial area and liquid-side mass transfer coefficient.

Mass transfer coefficient in the gas phase, k_G, is modeled by the dimensionless equations derived from the theoretical considerations and experiments with SO_2 chemisorption from helium, nitrogen and SF_6 and valid also for NH_3 absorption from the air. The gas Schmidt number range of Sc_G 0.49 - 2.32 covered by experiments includes that of ordinary distillations. We assume the model valid also for the distillation conditions due to the relatively simple nature of the gas phase and broad range of the experimental conditions used for its development. Also the latest distillation experimental results indicate the transferability.

The distillation experiments with neutral, positive, and negative system revealed that the interfacial area for the positive system (n-hexane/cyclohexane) is 17 – 20% higher than that for chemically similar but negative system (cyclohexane/n-heptane), which is attributed to the Marangoni effects. For neutral system (ethanol/n-propanol) the values of the effective mass-transfer area agree with those obtained by chemisorption experiments. The effective interfacial area model is therefore based on the regression of the chemisorption experiments results and incorporates correction for the Marangoni effects.

Dimensionless equation for the mass transfer coefficient in the liquid phase, k_L, is derived from the experiments with oxygen desorption from water, methanol, ethanol, n-propanol. However, analysis of the composition profiles revealed negligible mass-transfer resistance of the liquid phase for the negative and positive system, while for the neutral system the liquid phase resistance is clearly detectable and $k_L a$ values reach 180 – 190% of those following from the desorption correlation. The k_L model is therefore applied for the systems with stabilization index close to unity (neutral systems), otherwise the liquid-side mass-transfer resistance is considered negligible.

The model assumes plug flow for the gas phase as air-water experiments did not show any significant deviations from the plug flow for the ordinary hydrodynamic conditions. The plug flow with axial dispersion is introduced for the liquid phase hydrodynamics. The corresponding values of the Bodenstein number are derived from the results obtained with air-water system as well as of distillation experiments.

The specific parameters of the here presented model are derived from the experimental data acquired with i.d. 300 mm distillation and i.d. 150/300 mm absorption columns for Mellapak packings.

640360 Sieve Tray Efficiency Models: A Comparative Study Using FRI Data

Abstract:

Tray efficiency is an important variable in the design of new columns or the retrofit of existing columns. Over the years, several efficiency correlations have been presented in the literature. They ranged from fundamental models to more empirical models that were based on field data. Many of the fundamental models were based on the original AICHE model, where improvements were made to some of the drawbacks in the older model. The more recent development in fundamental models involves the application of bubbling and jetting zones on an operating tray. The newer models include the bimodal distribution of bubbles into small and large bubbles and the contribution of the jetting fraction to the overall froth. These models provide a continuous transition from spray to froth regime so that separate models are not needed for the two regimes. In contrast, the empirical models widely used by industry practitioners were developed from field data and relate tray efficiency to physical properties such as liquid viscosity and relative volatility. In this study, we present a comparison of different tray efficiency models and investigate their relative performance using FRI measured data on sieve tray efficiencies. An analysis of the key results from the study will be presented.

639856 Negative Inlet Downcomer Seal Impact On Tray Efficiency: A Study of Past Commercial Data

Abstract:

Downcomer (DC) hydraulics form an integral part to the successful efficient operation of distillation cross flow sieve and valve trays.

Conventional wisdom for tray designers is to ensure downcomer (DC) seal liquid to facilitate stable liquid entry on to the tray deck. DC seal depends on tray designs including the DC clearance and outlet weir height and operating conditions. Parameters including DC backup, tray pressure drop, and tray liquid holdup all influence the DC seal. Downcomer backup (clear liquid height at bottom of DC) comprises of clear liquid height at DC exit, DC head loss, tray clear liquid height, outlet weir height.

Under certain conditions, when downcomer seal is lost, vapor short circuits through the downcomer, depriving the liquid passage through the downcomer on to the tray deck. A negative inlet downcomer seal is detrimental to overall column efficiency, particularly at turndown.

The paper comprises of two parts. Part 1 is a forensic analysis of four tray clear liquid height predictions from the literature compared against measured old FRI data in reports held in a special section of Oklahoma State University. When plugged into the clear liquid height at bottom of DC correlation, one can determine which literature correlations predict a positive or negative seal, particularly at turndown. Finally, it will be shown whether a negative DC seal deleteriously affects measured tray efficiencies documented in those same FRI reports.

639970 High Performance Flexipro® Fixed Valve Tray

Abstract:

The quest for energy savings, reductions in greenhouse gas emissions and maximizing return on investments are driving the development of distillation trays that maximize capacity and efficiency, while also considering reliability. Fixed valve trays offer mechanical reliability and fouling resistance benefits relative to moving valve trays. Moving valve trays have the advantage of maintaining reasonable efficiencies to lower vapor rates, which yields a wider operating range than conventional fixed valves or sieve decks. The FLEXIPRO® high performance fixed valve was developed to simultaneously optimize the capacity, efficiency, pressure drop, fouling resistance and reliability of valve trays. The tray has higher capacity than other fixed valves, turn-down ratios of larger than 4:1 were achieved, the efficiency is comparable with smaller fixed valves, larger clearances are used to address fouling concerns, and with it being a fixed valve, the mechanical reliability concern around moving valves is addressed. Experimental data from two large diameter pilot columns and examples of applications will be presented.

639809 Distillation in Ethylene Plants – an Overview

Abstract:

Distillation towers are major components of an ethylene plant. Located downstream the pyrolysis furnaces, there purposes can be divided in 3 categories: heat transfer, absorption, and fractionation. The operation functionality and design characteristics are driven by the service in which they operate and the defined process objectives. At a high level, the ultimate objective is to remove contaminants/ impurities and recover value chemicals to purities meeting saleable products. Problems associated with distillation tower operation typically involved fouling, controls and startup difficulties. This paper discusses functional characteristics, internals, and typical problems by service. The sensitivity to feedstock variations is also discussed.

640343 Natural Gas Cryogenic Demethanizer Troubleshooting:an Internal Look of the Column Operation

Abstract:

The Demathanizer column of a new natural gas plant located in the West Texas area showed symptoms of premature flooding after one year of uninterrupted operation. The performance issues started after the operation mode was switched from rejection to recovery. Under recovery mode the plant was not able to achieve its nameplate capacity (200 MMSCFD), because the Demethanizer column showed signs of flooding and excessive entrainment (high pressure drop, erratic temperature profile, out of spec products, etc.) while operating with the lower temperature profile and higher liquid rates required for this mode of operation.

A step-by-step analysis was conducted to identify the possible causes of the performance issues. Throughout the analysis different tools and techniques were used to aid with the interpretation of the process data and column behavior: process simulation, hydraulic evaluation, gamma scanning, on-site inspection of the column internals, etc. This presentation will provide an overview of the steps taken to complete the troubleshooting analysis and highlight some recommendations that could assist other engineers to interpret the process data of columns with similar performance.

640169 Troubleshooting Tower Internal Foulings

Abstract:

Tower internals fouling potentials are often overlooked during the design stage. When combined with poor maintenance it can result in capacity shortfall and shorter run lengths.

Case Study #1: A MTBE unit was converted to ETBE production. Since the conversion, ETBE's catalytic distillation column could not meet the design capacity due premature flooding. Catalytic section trays fouling was suspected as a root cause. As a corrective measure, the licensor recommended certain trays modifications in the catalytic section. However, after the tray modifications and cleanings, the column capacity did not improve. Subsequently the catalysts were completely removed to avoid trays from fouling. Column capacity improved, but premature flooding problem persisted. Thorough column inspection revealed heavy fouling of vapor channels and internal piping that were overlooked previously. After cleaning of all internals, the column design capacity was fully restored.

Case Study #2: Heavy aromatic compounds are purified in a series of distillation columns equipped with high-performance structured packing at elevated temperature and under vacuum. High temperatures and air leakage resulted in coking of the processed hydrocarbons. Coke formation during operation gradually plugged the liquid distributors as well as the structured packings. Ultimately the unit had to be shut down periodically for cleanings with average run lengths of only 3~4 months. Using columns inspection and a series of specialized performance tests, the root cause was attributed to poor liquid distributor designs. Liquid distributors were replaced with anti-fouling & easy maintenance design features. Results from the revamp are: (1) unit run length more than doubled (2) improved product qualities.

639249 Capacity of High Performance Trays Limited By Column Inlet Arrangement and Vapor-Liquid Physical Separation Behavior in an Industrial Distillation Tower

Abstract:

Often in distillation towers, focus is geared toward the tray or packing hydraulics and majority of optimization efforts focuses on improving the tower performance by optimizing trays / packings hydraulics. More distillation towers are equipped with high capacity internals every day and as the feed rates and column throughputs are increased, and attention is forced more toward the contacting devices and away from other aspects, hence, some basic fundamentals are neglected. This paper will review a few of those fundamentals about gas-liquid physical separation in a distillation tower and how an improper feed inlet arrangement design, limits the column capacity equipped with high capacity trays and floods the trays. The paper will review the actual industrial tower troubleshooting and provides solution and guidelines to check and resolve the column's performance bottleneck.

Gas Entrainment by Liquid Jets and Cascades Can Unsettle Towers

By

Henry Z. Kister, Garry Jacobs, and Abraham A. Kister

Fluor, Aliso Viejo, CA, USA

Presented at the Distillation Symposium, AIChE Spring Meeting, San Antonio, Texas, April 11-14, 2022

Gas Entrainment by Liquid Jets and Cascades Can Unsettle Towers

Henry Z. Kister [a], Garry Jacobs [a], Abraham A. Kister [a]

[a] Fluor,3 Polaris Way, Aliso Viejo, CA, 92698, USA
henry.kister@fluor.com; garry.jacobs@fluor.com; avi.kister@gmail.com

Abstract. Liquid jets that strike liquid surfaces entrain gas into the liquid. Normally, the gas can vent out with few adverse consequences. Entrained gas unable to freely vent out may cause liquid backup, instability, liquid maldistribution to packed towers, and excessive venting downstream.

Tests in scaled-down models of parting boxes and seal pans provided improved understanding of gas entrainment. Liquid jets issuing from spargers generate uneven frothing in packed tower parting boxes, which leads to liquid maldistribution upon turndown. Impingement plates can alleviate this maldistribution, but need to be vented to avoid trapping gas underneath. Gas entrainment may also be troublesome when liquid from seal pans drops 2 meters or more. When sheets of high surface tension liquid descend from seal pans, they tend to form narrow jets that may locally drag gas a considerable depth into the liquid below. With foaming liquids, the gas spreads out over a wide region.

Keywords Gas trapping, Cascades, Gas entrainment, Waterfall pool effect, Sump aeration, Parting box aeration, Venting, Foaming

When liquid jets strike a liquid surface, they entrain gas into the liquid. Most of the time the gas can vent out with no apparent consequences. When the entrained gas cannot freely vent out, it may cause liquid backup, instability, liquid maldistribution to packed towers, and excessive venting problems downstream, as in the situations described below.

Trapping of gas generated by liquid jets in packing distributor parting boxes

A parting box is a rectangular box typically mounted above packed-tower distributor troughs (Figure 1). A feed sparger pipe elevated just above the box and oriented parallel to it supplies liquid to the box. The pipe may contain holes in its underside or dip tubes that transport the liquid to near the bottom of the box[1]. The liquid may be overhead reflux, collected internal liquid mixed with liquid feed, or pumparound return liquid. Holes in the floor of the parting box, located above each of the troughs, meter the liquid to each trough (Figure 1), targeting a uniform liquid head in all the troughs.

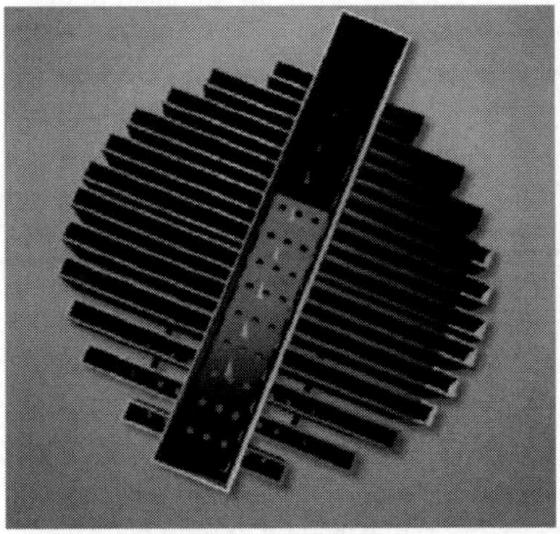

Figure 1 Parting box feeding trough distributor (Courtesy of Fractionation Research Inc.)

Figure 2 Sparger-fed parting box, showing uneven frothing

When feeding the box from a sparger containing holes in the underside (no dip tubes), the liquid jets issuing from the sparger holes impact the liquid surface in the box, generating intense uneven frothing in the box (Figure 2). This frothing leads to maldistribution of liquid to the troughs at turndown, as demonstrated by our extensive water tests of a scaled-down model of a parting box[1]. The froth height often rises above the liquid height predicted by the Bernoulli equation and may exceed the trough height leading to liquid overflowing the box, causing gross maldistribution.

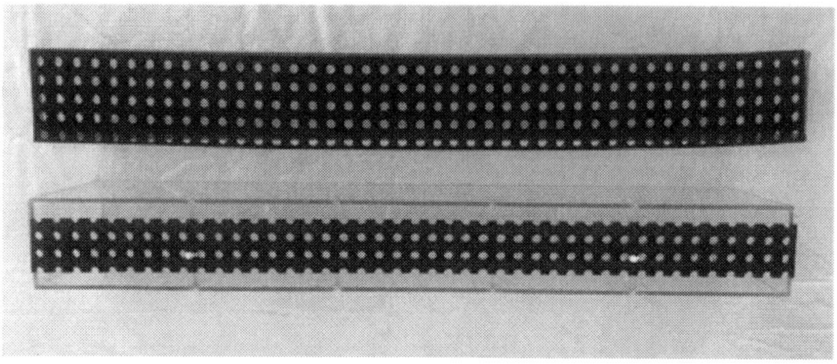

Figure 3 Perforated impingement plates. Top: Perforated impingement plate (PIP), same width as box. Bot: Vented impingement plate, has gaps between plate and box walls

A Perforated Impingement Plate (PIP, Figure 3 top), stretching wall-to-wall in the box, mounted 2-4" above the floor, with perforation area typically 10-20% of the box cross section area, is often used to make the froth uniform. The hole area of the PIP significantly exceeds the hole area at the bottom of the parting box, so the PIP does not impede the flow, just breaks the momentum, evens the froth, and improves the distribution of liquid exiting the box[1]. Our recent tests[1] identified a venting issue with PIP's. Water jets exiting the sparger feed pipe entrained air upon impact with the PIP and the water above it (Figure 4). This air passed downward through the PIP and was trapped beneath it by the water head above the PIP. This air gap generated excessive froth heights above the PIP, leading to premature overflow from the box. Overflowing liquid is maldistributed and often picked up by the vapor causing entrainment and premature flooding in packed towers, and is one of the prime causes of packed tower malfunctions[2].

Figure 4. Tests with 20% open area PIP, 100 mm above the parting box floor. The liquid level at the bottom of the box is only about 25 mm, with an air gap occupying the other 75 mm between the PIP and the floor of the box.

Adequate venting of the PIP is achieved in some commercial designs by using a narrower impingement plate that leaves gaps on the sides to permit gas venting. Such vented impingement plates (VIPs, Figure 3 bottom) were tested in the same test rig. The tests showed that using the VIP's with 40-50% open area completely eliminated the air gap, lowered the froth heights and eliminated the box overflow, while still providing an excellent liquid split to the parting box perforations at full rates and down to 50% turndown. The takeaway is that when perforated impingement plates are used in parting boxes, they should be well-vented. We have seen an operating tower where overflow from a parting box containing an unvented PIP caused a separation deterioration at high rates.

Gas entrainment due to cascades in sumps

Gas entrainment due to cascading ("waterfall pool effects") is caused by the impact of falling liquid on the liquid surface in the bottom sump, draw pan, or reflux drum. The impact induces gas entrainment into the liquid in the same manner as a waterfall induces air entrainment into the pool beneath it. Both the quantity of gas entrained and the depth to which the gas bubbles penetrate below the liquid surface are enhanced when large quantities of liquid fall into a small area in the sump, or when the vertical drop height is large. This gas entrainment is a much greater problem in tray tower sumps, where the liquid fall resembles a waterfall, than in packed tower sumps, where the liquid fall is like rain. With foaming systems this problem is compounded, as the waterfall generates foam and the gas bubbles are difficult to disengage from the foam. This is analogous to water descending from a tap generating foam in a dish containing detergent.

Although this mechanism has been known for decades[3], to the best of our knowledge, the only industry experience reported with it was in a vertical 0.76 m diameter reflux drum[4] of a chemical tower. Liquid feed, entering near the top of the drum (Figure 5), dropped about 2 m into a shallow pool of liquid, about 0.5 m above the liquid outlet nozzle at the bottom of the drum. The impact of the falling liquid onto the shallow pool entrained vapor bubbles, choking the outlet line that was undersized for self-venting flow[5, 6]. Attempts to raise the liquid level led to intermittent degassing of the drum liquid. The degassed liquid siphoned out of the drum, causing severe instability. The solution was rerouting the feed to a lower nozzle (Figure 5), which eliminated the cascade and restored stability.

In an unpublished case study by one of the authors, instability that initiated at a chimney tray side draw unsettled a refinery fractionator. Here, the cascade over the chimney tray weir (Figure 6), re-entrained vapor into the side draw takeoff sump. Similar to the reflux drum above, the draw line was undersized for self-venting flow[5, 6], and was choked by the entrained vapor bubbles, backing liquid up the draw sump. When enough liquid backed up in the draw sump, the waterfall height diminished, vapor entrainment stopped, and the liquid siphoned out. The process would then start again.

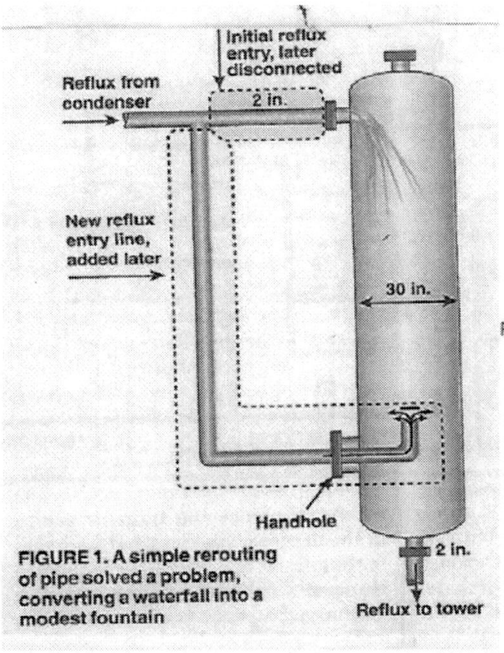

(from Ref. 4, reprinted courtesy of Chemical Engineering)

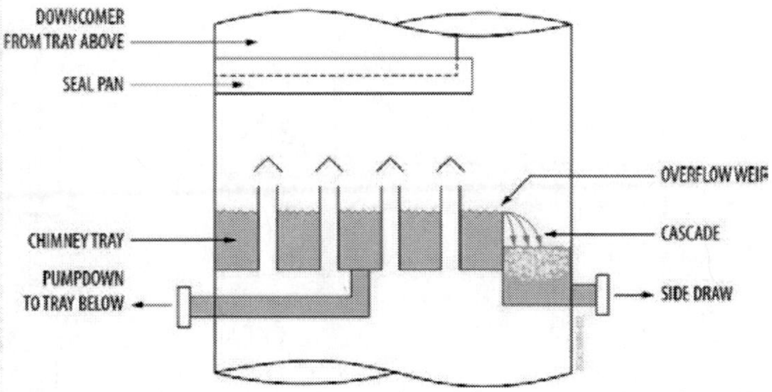

Fig.6. Cascade aerating chimney tray outlet draw liquid

In another unpublished case[8], liquid reflux dropped 12 inches from the feed pipe to the liquid sump behind the inlet weir (Figure 7a). The tower experienced foaming. A gamma scan showed foam on a few trays, but a real bad foam only on the top tray. To cure, the pipe was extended so that its outlet became submerged below the liquid level (Figure 7b). No more foaming after this.

Another foaming experience took place in a regenerator of a hot pot absorber. The liquid entered via a pipe distributor with holes pointing downwards into a sump behind an inlet weir (Figure 7c). Again, gamma scans showed foam on the top tray. To cure, the pipe was fitted with dip tubes that were submerged by the sump liquid (Figure 7d). No more foaming after this.

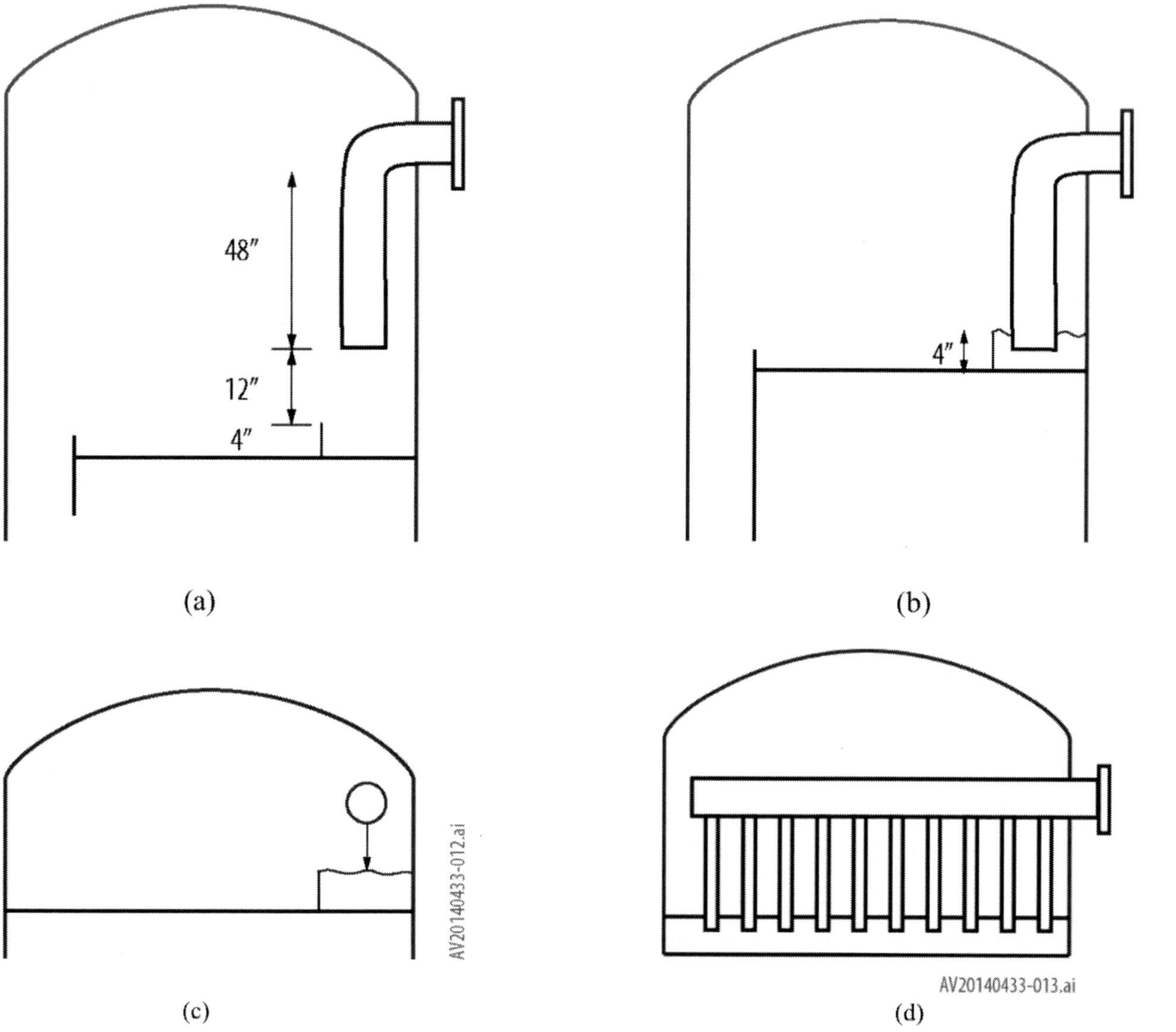

Figure 7 Reflux entry into towers with foaming systems (a) Dropping the reflux from a pipe 12 inches into a liquid pool produced tower foaming[8] (b) Extending the pipe into the pool eliminated the foaming[8] (c) Another foaming system, reflux introduced via a pipe distributor with holes pointing downwards into the pool, foaming produced (d) Foaming eliminated by adding dip tubes that brought the reflux below the sump liquid level

Gas entrainment due to cascades in bottom sumps

One place where gas entrainment can be most troublesome is in the bottom product. Besides causing instability, choke, and bottom pump cavitation, gas entrainment can also cause severe venting problems and environmental issues downstream. While the high residence times usually incorporated in the design of bottom sumps[3] are highly effective in degassing entrained bubbles, there are times, especially with foaming systems, that the cascade-generated entrained gas still finds its way into the bottom draw. Le Grange et. al.[7] reported severe carry-under of gas with the rich amine leaving an absorber into the downstream rich amine flash drum. The excessive gas in the flash drum resulted in high flaring of product gas (about 3%) and frequent safety valve lifting. The cause was foaming due to a contaminating aqueous phase surfactant entering with the feed gas.

We have experienced a similar problem in a chemical tower handling a foaming system. Liquid cascade induced excessive downstream venting due to carry-under of gas into the bottom stream. Due to the large diameter of the tower, the transmission of gamma rays through the bottom sump was too low to identify significant aeration in the sump. In our troubleshooting endeavors, we performed some backyard qualitative tests in which we dropped water into a tank to simulate the cascade into a tower bottom sump. In these tests, we placed a 75-liter, 0.4 m tall fish tank at ground level and filled it to the top with water to simulate the bottom sump. We mounted a second 40 liters tank some height above the first tank, filled it up with water, then tilted it so water flowed over its narrow edge (250 mm) and cascaded onto the lower 75 liter tank. We timed the emptying out from start to finish by a stop watch, which enabled calculating the liquid flow rate. This modelled the seal pan. The tilt was manually raised during each run to keep a constant weir load of about 40 m3/h m of length, similar to the weir load over the seal pan lip in the tower we were troubleshooting. The tilt of the tank was manually adjusted by lifting the opposite end of the tank, and the emptying out lasted about 10-12 seconds. The hand-lifting was obviously a very crude means of flow control and the flow was not all that uniform during the run. This was not meant to be a precise test. The idea was to learn more about the physics of the cascade.

In the first two tests, the drop height was 2.2 meters, which is a typical cascade height in bottom sumps. In the first test, the 75 liter tank contained water. In the second test, dishwasher detergent was added to the water in the 75 liter tank to investigate cascade into a sump containing a foaming solution. The final two tests repeated the initial two tests, but with a cascade of only 0.8 meters.

The 2.2 meters elevation drop tests, Figure 8, showed the following:

1. The liquid sheet leaving the upper tank, initially 250 mm wide, narrowed down to about 50 mm width during the fall. The 50 mm jet dragged air with it all the way to the bottom of the tank (0.4 m deep).
2. The width of the jet hitting the water was not uniform, ranging from 25 mm to about 100 mm wide. During moments when the liquid rate was larger due to excessive tilt, the jet was wider, and penetrated to a smaller depth (200-300 mm),
3. With no detergent in the bottom tank (Figure 8a), the aeration was local, near the point where the jet impacted the water, with only a slight horizontal spread. In contrast, with the detergent in the lower tank (Figure 8b), once the jet caused foaming, the foam spread right through the tank and took some time to clear.

The 0.8 m elevation drop tests, Figure 9, showed the following:

4. The liquid sheet leaving the upper tank, initially 250 mm wide, narrowed down to a lesser extent, to around 150 mm. The jet dragged air with it but only to a depth of 150-300 mm with no detergent (Figure 9a), but deeper when the detergent was present (Figure 9b).
5. The width of the jet hitting the water was not uniform, ranging from 100 mm to about 200 mm wide. During moments when the liquid rate was larger due to excessive tilt, the jet was wider, and penetrated to a smaller depth,
6. Same as 3 above, but with Figure 9.

The reason for the narrowing width of the falling water sheet is gravitational acceleration. Exiting the tank, the water had no downward velocity, but accelerated as it fell toward the tank below. The increasing velocity narrowed the cross-sectional area of the falling water. By the principle of continuity, the velocity times area needs to remain constant, so as the water sped up, the flow area diminished. The more the water accelerated, the smaller the area, narrowing the sheet width. The high surface tension of the water keeps the water sheet together. With the 2.2 m drop, the velocity rise due to the high free fall turned the sheet into a water jet. With the shorter 0.8 m drop, the velocity was lower at the point of impact and the sheet width remained larger. With the 2.2 m drop, the high-velocity jet penetrated the liquid over a small area, dragging the gas bubbles further down. The depth of penetration would be directly related to the velocity head as it hits the water surface.

A similar mechanism is likely to have intensified the problem in the tower in Figure 7a. The reflux pipe was most likely not running liquid full, because the liquid begins to accelerate, due to gravity, when in passes through the elbow. As a result, its initial velocity (in the horizontal direction) increases as it falls 60 inches into the sump behind the inlet weir. Continuity is preserved because vapor backflows into the reflux pipe.

Discussion

Our tests provided insight into the cascading phenomenon. They showed that liquid cascading about 2.2 m tends to narrow into a much smaller width, forming an energetic liquid jet that drags gas bubbles a considerable distance at the location of impact. When the cascade height is shorter, about 0.8 m, there is insufficient height for the cascade to narrow, causing the liquid to land in the sump like a sheet, resulting in the gas bubbles penetrating to a smaller depth. Presence of foaming materials in the sump leads to much wider, and possibly deeper, penetration of gas bubbles into the sump.

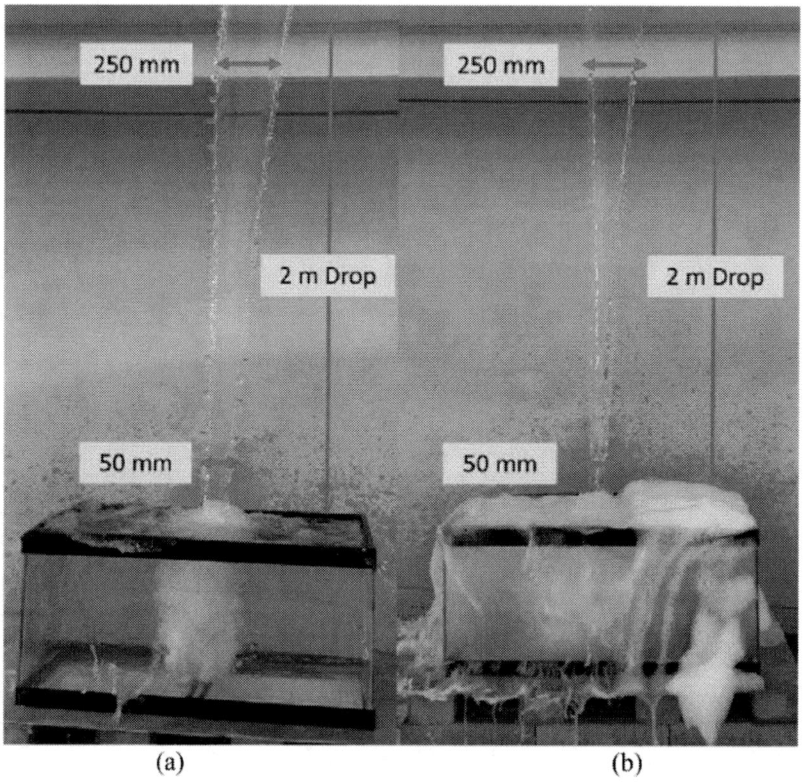

Figure 8. Dropping water sheet about 250 mm wide from about 2.2 m into a tank. (a) water in tank (b) water with dishwashing detergent in the tank, simulating foaming material in sump

The narrowing of a descending sheet into a jet is dependent on the surface tension. In our tests, the high surface tension held the water together, causing the wide sheets to deform into narrow jets. With low surface tension systems, the sheets are more likely to break into smaller, less energetic streams or droplets, so the aeration induced by the cascade will be reduced.

Both the systems that we were troubleshooting and that reported by le Grange et. al.[7] are high surface tensions systems, close to that of water, and are foaming systems. Our tests showed that in foaming systems gas entrainment due to cascading is much more severe than with non-foaming systems. In the system we were troubleshooting, injection of a few ppm antifoam mitigated the gas carry-under and venting. After stopping the antifoam, the gas venting rate rose, falling again as soon as the antifoam injection was resumed. In the case of le Grange et.al.[7], eliminating the foaming component from the tower feed eliminated the gas carry-under and flaring.

(a) (b)

Figure 9. Dropping water sheet about 250 wide from about 0.8 m into a tank. Water sheet narrows a lot less, which upon impact on the tank water level induces much less gas entrainment that penetrates lesser depth into the tank liquid (a) water in tank (b) water with dishwashing detergent in the tank, simulating foaming material in sump.

In the tower we were troubleshooting we also attempted to eliminate the need for antifoam by reducing the cascade fall height from 2 m to 1 m. This was not successful, probably because about half the liquid came from the feed nozzle, and the drop height of this liquid remained 1.7 m, generating liquid jets similar to those in Figure 5.

Takeaways

Gas is entrained into a liquid pool when liquid jets or cascading liquid sheets impact the pool liquid surfaces, dragging gas bubbles into the pool. This aeration can generate uneven froth in parting boxes or bottleneck draw lines, and needs to be considered in the design and troubleshooting of distillation equipment.

Our tests showed that when liquid sheets of high-surface tension liquid descend, they narrow in width, turning the sheet into a high-velocity liquid jet that can penetrate and drag gas a considerable distance into the pool. In most cases, the large residence times for which bottom sumps are designed for will prevent gas entrainment into the bottom outlet. Our tests show that with non-foaming liquids, the aeration remains local, near the point where the jet impacts the liquid, with only a slight lateral spread. In contrast, with foaming systems, the aeration spreads over a wide area. This can induce significant gas entrainment into the bottom outlet in foaming high-surface tension systems, especially when the liquid cascades from heights of 2 meters or more. This gas entrainment can lead to excessive venting downstream with product losses and possible need for flaring. It can also lead to instability and to cavitation of the bottom pump. Eliminating the foaming and restricting the liquid fall height are possible cures.

References

1. H. Z. Kister, G. Jacobs, and A. A. Kister, *Chem. Eng. Progr.*, May, p.22, 2021
2. H. Z. Kister, *Chem. Eng. Res. Des.*, 81, Part A, p. 5, January 2003.

3. H. Z. Kister, *Distillation Operation*, McGraw-Hill, NY, 1990.
4. H. Z. Kister and J. F. Litchfield, *Chem. Eng.*, p.56, September 2004
5. L. L. Simpson, *Chem. Eng.*, June 17, p. 192, 1968.
6. A. Sewell, *The Chem. Engnr.*, 299/300, p. 442, 1975.
7. P. le Grange, M. Sheilan, and B. Spooner, *Sulphur* 368, p. 42, January – February 2017
8. Olsson, F. R, Private communication, 2019.

AUTHOR INDEX

AIChE
120 Wall Street, FL 23
New York, NY 10005-4020

CURRAN ASSOCIATES INC.
proceedings
.com

ISBN 978-1-7138-6820-0
9 781713 868200